OCEAN ENGINEERING STRUCTURES

Volume I

B 94

(Course Notes)

J. Harvey Evans

and

John C. Adamchak

THE M.I.T. PRESS

Massachusetts Institute of Technology

Cambridge, Massachusetts and London, England

69-18

Second Printing, December 1970

ISBN 0 262 55001 6 (paperback)

Library of Congress catalog card number: 77-93040

Printed in the United States of America

MASSACHUSETTS INSTITUTE OF TECHNOLOGY

DEPARTMENT OF NAVAL ARCHITECTURE

AND MARINE ENGINEERING

CAMBRIDGE, MASS. 02139

ACKNOWLEDGMENT

This text was prepared under the auspices of the
Department of Naval Architecture and Marine Engineering
at Massachusetts Institute of Technology. Its preparation
was supported in part by a grand made to M.I.T. by the
Ford Foundation for the purpose of aiding the improvement
of engineering education, in part by a grant from the
National Science Foundation under the terms of the Sea
Grant Project No. GH-1 for Curriculum Development for Ocean
Engineering Education, and in part by M.I.T. funds.

The material contained in this volume was developed
and written by Professor J. H. Evans and Mr. John C.
Adamchak for use in a subject entitled "Ocean Engineering
Structures" which is offered by the Department of Naval
Architecture and Marine Engineering as part of its Ocean
Engineering Program.

Alfred H. Keil, Chairman

May, 1969

During the Fall Term of 1967 a subject designated "Ocean Engineering Structures" was given for the first time at M.I.T. It was developed as one of the basic, core areas to be included in the newly established graduate Ocean Engineering curriculum administered by the Department of Naval Architecture and Marine Engineering. As such, the subject was intended to be introductory and cater to the varied backgrounds of a wide range of engineering and science graduates. At the same time, the subject matter was to be treated in a meaningful way, even for those with a considerable depth of exposure in structural analysis and design. On the whole, these diverse objectives appear to have been fairly well met and the text which follows is a concentration of the material covered, or in fact an outline in narrative form.

On this first shakedown trial the subject was conducted largely on a seminar plan with each of the 25 students being responsible for both written and oral presentations (one hour long) on an assigned subject. The remaining 20 sessions were occupied with a sustained sequence of lectures on cylindrical and spherical shells under hydrostatic pressure; cylindrical shells under axial, bending and complex loadings; and on the design of stiffened cylindrical shells, given by faculty. Several individuals with outstanding familiarity with one or another of the subjects were also brought in for special lectures. As it turned out, several members of the class were themselves extremely well qualified to speak knowledgebly in certain areas of coverage.

Each headed section of the text with follows represents the material in one 50 minute session (except as indicated) with a small amount of associated discussion understood. It is based on the detailed subject outline given to students at the beginning of the semester. By now it represents the edited output of many sources as well as numerous original contributions by the authors. The brief introductory treatment of the ocean environment is in no sense complete, but is intended to highlight the opportunities for productive and beneficial occupation in the oceans and to establish the boundary conditions within which ocean engineering is to be carried on. Contained within the first three sections of these notes are a number of direct quotations which seem particularly cogent from a variety of authoritative sources. They have not been specifically acknowledged in each case because of the random manner in which they have been inserted. Foremost among the sources, however, were the books, "The Earth Beneath the Sea" and "Submarine Geology" by F. P. Shepard; either of which might serve as a text.

For present purposes, ocean engineering structures are grouped as line structures (including moorings),

surface platforms and submarine pressure vessels. The
omission of surface vehicles, whether of the positive
displacement (including catamarans), hydrofoil, hydroplane,
or air cushion type, is on the grounds of their being
included already and thoroughly in other well established
naval architecture subjects. The same reasoning applies to
the omission of any more than introductory, interface
coverage of space frame analysis, soil mechanics and materials,
regardless of how deficient other curricula or subjects
might be in dealing adequately with the particular aspects
of importance in the deep ocean environment. The roles of
the strength and mass density properties of materials in
tradeoff analyses are covered, however, especially for
their importance in submarine pressure hull design.

The classifications of source references found in
the Appendix are not altogether coincident with the seminar
subject titles. This arose in attempting to avoid excessive
duplications of listing. A few of the references will be
found more of news value than technical. In general, this
is the case with significant new developments not yet documented
in any other form and is indicative of rapid change. In
most instances, professional papers would subsequently displace
the news items. Periodicals cited include those received
through December 1, 1968. For convenience, the reference lists
are in Volume II of these notes.

Present thinking indicates continued adjustment
and shaping of this structures subject within the context
of the M.I.T. Ocean Engineering curriculum. It must
continue to serve as a self contained and terminal subject
for those requiring an introduction to the field. Probably
it will also be called upon to function as a foundation
for those wishing to specialize in structures and for whom
a second subject in deep ocean structural analysis and
design will be developed, treating certain special topics
in greater depth.

OCEAN ENGINEERING STRUCTURES

Contents

THE DEEP OCEAN ENVIRONMENT

Winds and Waves

In water depths of less than 100 feet, wind loads
on exposed stationary structures are likely to be those most
critical. In areas of hurricane occurence, winds with maximum velocities of 125-135 miles per hour can be expected as,
for example, with hurricane "Carla" which struck the Gulf of
Mexico in 1961 bringing hurricane force winds to distances of
100-120 miles out from its center. Statistically, such storms
are the worst likely to occur in a 100 year period. Nevertheless, for the Gulf region, they are now the usual basis
for design, although design winds may be reduced to 100 miles
per hour when in combination with maximum design wave conditions.

The 'maximum velocity" has been defined as the greatest
average velocity for a five minute period. The "extreme velocity" is the velocity for the fastest mile and is about 20%
greater than the maximum velocity. Gust velocities are usually
taken to be 20% higher than the extreme velocity. Wind velocities in the boundary layer increase with the 1/7 power of
distance above the water surface.

Several twenty-five year storms may occur in the
Gulf within one year; but at various locations. Among dispersed,
less costly offshore structures formerly built to 25 year storm
criteria, random losses were quite frequent. On the other hand,
with a compact grouping, losses might be less frequent but
involve more structures and so be more sizable. In high risk

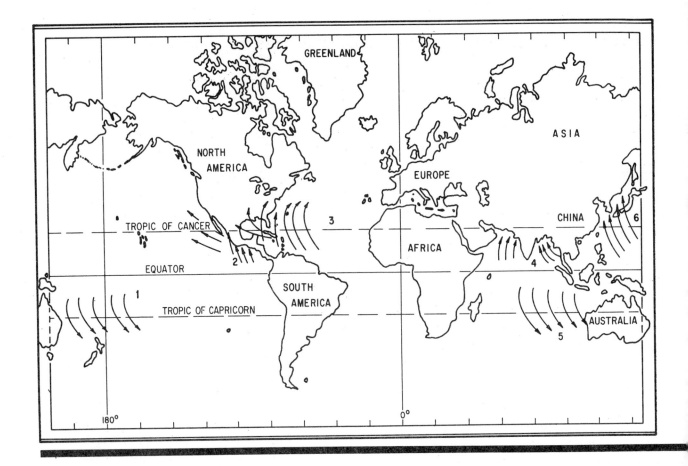

Tropical storms occur in six regions of the earth as shown on map ▰▰▰

1. In the region east of Australia in the South Pacific, the storms are called "cyclones," and sometimes "typhoons" or "willie-willies." They can be expected in any month of the year, but the greatest frequency comes in the January-to-March period, inclusive.

2. Off the west coast of Mexico, the storms are usually called "tropical cyclones" or "hurricanes" and they occur in the period starting with June and ending with October.

3. In this region, the storms are almost invariably referred to as "hurricanes" and they occur in the six-month period starting with June and ending with November. Greatest frequency is in August, September and October.

4. In this region, on either side of India, the storms are properly called "tropical cyclones." They occur in all months except February with the two periods of highest frequency coming in May and October.

5. In this area south of the equator, in the South Indian Ocean, "cyclones" can be expected in all months except July and August. The period of greatest frequency comes in December through March.

6. In the China Sea and North Pacific Ocean the storms are invariably called "typhoons." These great storms occur in the period from May to December with the greatest frequency coming in July, August and September. These storms are usually larger than their counterparts in other parts of the earth.

It should be noted that tropical storms do not occur in the South Atlantic Ocean.

Hurricane and typhoon names

Area 3. The 1967 names of hurricanes for the Gulf of Mexico and North Atlantic are as follows: Arlene, Beulah, Chloe, Doria, Edith, Fern, Ginger, Heidi, Irene, Janice, Kristy, Laura, Margo, Nona, Orchid, Portia, Rachel, Sandra, Terese, Verna, Wallis.

NOTE: Where a hurricane seriously affects the U.S., the name assigned to it is retired for a period of 10 years.

Area 2. The 1967 names of tropical storms in the eastern Pacific off the west coast of Mexico are as follows: Agatha, Bridget, Carlotta, Denise, Eleanor, Francene, Georgette, Hilary, Ilsa, Jewel, Katrina, Lily, Monica, Nanette, Olivia, Priscilla, Ramona, Sharon, Terry, Veronica, Winifred.

Area 6. The entire list of names of typhoons in this area in the western Pacific are used regardless of the year in which they occur. The current list is as follows: Agnes, Bess, Carmen, Della, Elaine, Faye, Gloria, Hester, Irma, Judy, Kit, Lola, Mamie, Nina, Ora, Phyllis, Rita, Susan, Tess, Viola, Winnie, Alice, Betty, Cora, Doris, Elsie, Flossie, Grace, Helen, Ida, June, Kathy, Lorna, Marie, Nancy, Olga, Pamela, Ruby, Sally, Tilda, Violet, Wilda.

competitive ventures of this kind, the need is great for reliable, long range predictions of the environmental conditions peculiar to the location, defined in time and space.

Waves are wind generated and result from frictional drag on the water surface. The characteristics of a simple wind wave system are determined by the length of time the wind blows, the distance or surface area over which it blows (fetch) and its strength. After formation, waves move out of the generating area and may proceed up to hundreds of miles with gradually diminishing heights while being overlaid with other systems from other areas in the formation of the complex wave pattern of a "sea." It is therefore necessary in wave forecasting to consider storms occurring in all parts of the sea or ocean in which is located the site of interest.

At any given instant, at any particular location, ocean waves are usually not uniform but have a variety of heights and periods. The "significant height" is the average height of the highest 1/3 of all waves present and is about 1.6 times the average for all the incident waves. The maximum height will be about 1.8 times the significant height. As orders of magnitude for the Gulf of Mexico, and in 70-100 feet of water, the measured maximum wave height above mean low water in a 25 year storm was 32 feet. In a 100 year storm, it reached 42 feet.

Waves entering shallow water are transformed under the influence of bottom topography. When the wave crest

WIND AND SEA SCALE FOR FULLY ARISEN SEA

SEA STATE [1]	DESCRIPTION [2] (SEA-GENERAL)	(BEAUFORT) WIND FORCE	DESCRIPTION (WIND [3])	RANGE (KNOTS)	AVERAGE (KNOTS)	WAVE HEIGHT FEET — AVERAGE	SIGNIFICANT	1/10 HIGHEST	SIGNIFICANT RANGE OF PERIODS (SECONDS)	T_{max} PERIOD OF MAXIMUM ENERGY OF SPECTRUM	$\bar{T}$ (AVERAGE PERIOD)	AVERAGE WAVE LENGTH	MINIMUM FETCH (NAUTICAL MILES)	MINIMUM DURATION (HOURS)
0	Sea like a mirror.	0	Calm	Less than 1	0 [a]	0	0	0	–	–	–	–	–	–
	Ripples with the appearance of scales are formed, but without foam crests.	1	Light Airs	1 - 3	2	0.05	0.08	0.10	up to 1.2 sec	0.7	0.5	10 in.	5	18 min
1	Small wavelets, still short but more pronounced; crests have a glassy appearance, but do not break.	2	Light Breeze	4 - 6	5	0.18	0.29	0.37	0.4-2.8	2.0	1.4	6.7 ft	8	39 min
	Large wavelets, crests begin to break. Foam of glassy appearance. Perhaps scattered white horses.	3	Gentle Breeze	7 -10	8.5	0.6	1.0	1.2	0.8-5.0	3.4	2.4	20	9.8	1.7 hrs
				10	0.88	1.4	1.8	1.0-6.0	4	2.9	27	10	2.4
2				12	1.4	2.2	2.8	1.0-7.0	4.8	3.4	40	18	3.8
	small waves, becoming larger; fairly frequent white horses.	4	Moderate Breeze	11-16	13.5	1.8	2.9	3.7	1.4-7.6	5.4	3.9	52	24	4.8
3				14	2.0	3.3	4.2	1.5-7.8	5.6	4.0	59	28	5.2
				16	2.9	4.6	5.8	2.0-8.8	6.5	4.6	71	40	6.6
4	Moderate waves, taking a more pronounced long form; many white horses are formed. (Chance of some spray).	5	Fresh Breeze	17-21	18	3.8	6.1	7.8	2.5-10.0	7.2	5.1	90	55	8.3
				19	4.3	6.9	8.7	2.8-10.6	7.7	5.4	99	65	9.2
				20	5.0	8.0	10	3.0-11.1	8.1	5.7	111	75	10
5				22	6.4	10	13	3.4-12.2	8.9	6.3	134	100	12
	Large waves begin to form; the white foam crests are more extensive everywhere. (Probably some spray).	6	Strong Breeze	22-27	24	7.9	12	16	3.7-13.5	9.7	6.8	160	130	14
				24.5	8.2	13	17	3.8-13.6	9.9	7.0	164	140	15
6				26	9.6	15	20	4.0-14.5	10.5	7.4	188	180	17
	Sea heaps up and white foam from breaking waves begins to be blown in streaks along the direction of the wind. (Spindrift begins to be seen).	7	Moderate Gale	28-33	28	11	18	23	4.5-15.5	11.3	7.9	212	230	20
				30	14	22	28	4.7-16.7	12.1	8.6	250	280	23
				30.5	14	23	29	4.8-17.0	12.4	8.7	258	290	24
				32	16	26	33	5.0-17.5	12.9	9.1	285	340	27
7				34	19	30	38	5.5-18.5	13.6	9.7	322	420	30
	Moderately high waves of greater length; edges of crests break into spindrift. The foam is blown in well marked streaks along the direction of the wind. Spray affects visibility.	8	Fresh Gale	34-40	36	21	35	44	5.8-19.7	14.5	10.3	363	500	34
				37	23	37	46.7	6-20.5	14.9	10.5	376	530	37
				38	25	40	50	6.2-20.8	15.4	10.7	392	600	38
				40	28	45	58	6.5-21.7	16.1	11.4	444	710	42
8	High waves. Dense streaks of foam along the direction of the wind. Sea begins to roll. Visibility affected.	9	Strong Gale	41-47	42	31	50	64	7-23	17.0	12.0	492	830	47
				44	36	58	73	7-24.2	17.7	12.5	534	960	52
				46	40	64	81	7-25	18.6	13.1	590	1110	57
	Very high waves with long overhanging crests. The resulting foam is in great patches and is blown in dense white streaks along the direction of the wind. On the whole the surface of the sea takes a white appearance. The rolling of the sea becomes heavy and shock-like. Visibility is affected.	10	Whole Gale*	48-55	48	44	71	90	7.5-26	19.4	13.8	650	1250	63
				50	49	78	99	7.5-27	20.2	14.3	700	1420	69
9				51.5	52	83	106	8-28.2	20.8	14.7	736	1560	73
				52	54	87	110	8-28.5	21.0	14.8	750	1610	75
				54	59	95	121	8-29.5	21.8	15.4	810	1800	81
	Exceptionally high waves (Small and medium-sized ships might for a long time be lost to view behind the waves.) The sea is completely covered with long white patches of foam lying along the direction of the wind. Everywhere the edges of the wave crests are blown into froth. Visibility affected.	11	Storm*	56-63	56	64	103	130	8.5-31	22.6	16.3	910	2100	88
				59.5	73	116	148	10-32	24	17.0	985	2500	101
	Air filled with foam and spray. Sea completely white with driving spray; visibility very seriously affected.	12	Hurricane*	64-71	> 64	> 80 [b]	> 128 [b]	> 164 [b]	10-(35)	(26)	(18)	~	~	~

*For hurricane winds (and often whole gale and storm winds) required durations and fetches are rarely attained. Seas are therefore not fully arisen.

a) A heavy box around this value means that the values tabulated are at the center of the Beaufort range.

b) For such high winds, the seas are confused. The wave crests blow off, and the water and the air mix.

[1] Encyclopedia of Nautical Knowledge, W.A. McEwen and A.H. Lewis, Cornell Maritime Press, Cambridge, Maryland, 1953, p. 483

[2] Manual of Seamanship Volume II, Admiralty, London, H.M. Stationery Office, 1952, pp. 717-7

[3] Practical Methods for observing and forecasting Ocean Waves, Pierson, Neumann, James, N.Y. Univ, College of Engin, 1953.

Based on the Neumann Spectrum

This table compiled by Wilbur Marks, David Taylor Model Basin, 1956

advances over a shoaling bottom, refraction takes place, bending the crest as the deep water portions of the wave overtake those in shallower water. Finally, as the combination of wave height and water depth becomes less than the critical value, the wave breaks.

The force exerted upon a structure by waves is related to the partical or orbital motion of the waves. The dynamic pressures exerted by the tops of breaking waves represent the greatest pressures to which structures may be subjected by wave action because at the very crests of breaking waves the orbital velocity is at a maximum.

Techniques of wave forecasting involve calculation of deep water wave characteristics from wind-stress relationships. Such waves are then followed through their transformation by travel as swell from their point of origin, current effects, interference with other wave trains, bottom topography, bottom friction, breaking, secondary wave effects, etc. The two primary methods have involved analysis of wave trains through a characteristic wave height or period called the "significant wave" (Sverdrup, Munk and Bretschneider) or through wave spectrum analysis (Pierson, Neumann and James). The former has been described as the most practical, but the latter has potentialities of being most effective in large ocean areas where more complicated wave-swell systems develop.

A particularly dangerous feature of hurricanes is the storm tide raised by the dual effect of barometric suction

on the sea surface and the stress of the wind in piling up
water near the crest. The movement of the hurricane dynamically augments this rise in sea level above what could be
attained by steady winds or stationary pressure. Extreme
hurricane tides of over 10 feet have been reported along the
shore line of the Gulf of Mexico.

Internal waves operate within a body of water without
having any effect on the surface. While the height of such
waves is greatest at intermediate depths, the horizontal velocity is greatest nearest the surface and at the bottom where
it may be sufficiently powerful to disturb the ocean floor and
even produce ripple marks as do bottom currents.

Until very recently, ocean currents were thought to
be surface phenomena. However, currents at least comparable
to those of land rivers are now believed to develop from time
to time on the oceanic marginal slopes and to continue down to
the deep ocean floor.

THE DEEP OCEAN ENVIRONMENT

Currents, Thermal Structure and Ice

Oceanic currents such as the Gulf Stream are set in motion by the two sets of prevailing winds; the easterly trade winds in the tropics and the westerlies in the higher latitudes. Separate currents flow around the ocean basins north and south of the equator, in a clockwise direction in the Northern Hemisphere and in a counter clockwise direction in the Southern Hemisphere because of the prevailing direction of the winds. There is evidence that strong Gulf Stream currents continue down to depths of almost a mile, where they sweep the bottom.

In contrast, the Comwell or Pacific Equatorial undercurrent follows the Equator eastward for 3,500 and perhaps 8,000 miles. Although extending only from 50-100 meters to 300-350 meters below surface, it reaches out 125 miles on each side of the Equator and has core speeds of 2 1/2 - 3 knots.

With tidal currents, at least theoretically, the flow continues with little loss of velocity from the surface almost to the bottom so that bottom erosion due to tides is often very pronounced. The effects of these currents are strongest where constrictions such as straits or narrow entrances to large bays cause strong flows. Even in the open areas of the Gulf of Mexico, hurricane induced waves and bottom currents have moved unburied pipelines in 90 feet of water.

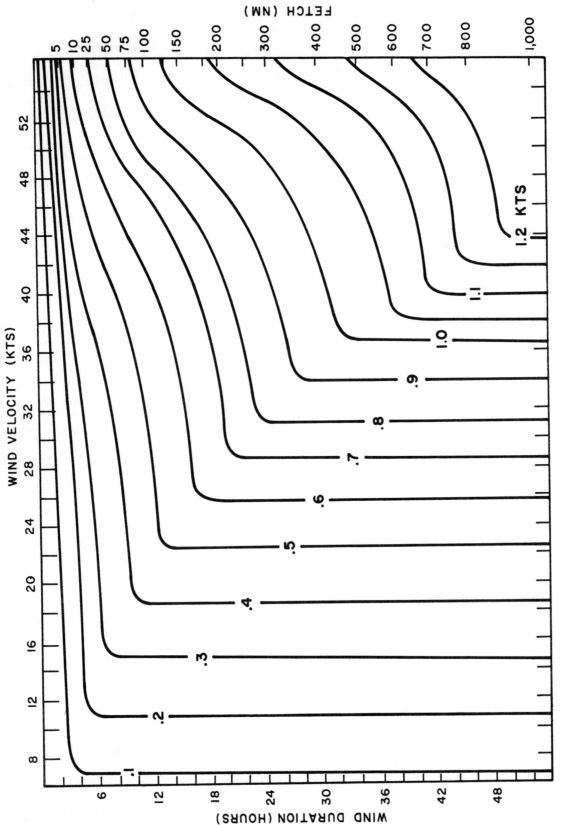

WIND DRIFT CURRENTS (Kts)

Enter with wind velocity and read drift current for the appropriate
wind duration or fetch, whichever gives the smaller current.
From "Ocean Thermal Structure Forecasting", U.S.N. Oceanographic Office, SP-105,
ASWEPS Manual Series, Vol. 5.

Mud stirred into the water will make the water heavier and hence the mud and water will flow down the slope. The possible existence of powerful "turbidity currents" of this type on the ocean floor is suspected. There are indications they are capable of transporting coarse sediment out into very deep water. They may also be important agents in eroding the ocean floor.

The thermal structure in the oceans is related to water masses and geographical location. The most irregular structures are found at oceanographic fronts and in major current systems. It has been found that the ocean is made up of many relatively thin layers increasing in density with depth. The layers may be sharply separated or they may diffuse into gradual gradients. The significant feature is that the thickness of the layers increases and decreases with distance and time. Distinguishable by their temperature, the layers are wave-like in shape and are in constant motion.

The temperature of water in the oceans decreases with depth. This gradient is accentuated toward the tropics where the difference in temperature between the surface and, say, 1,300 feet down can be on the order of 36°F. Beneath the Atlantic waters is a mass of cold bottom water (-0.8 to -1.0°C) that fills the submarine depressions.

In the Arctic, ice may have to be contended with in several forms; in sheets, rafted ice, perhaps many feet thick formed by one sheet slipping over or under another and as chunks or icebergs. In Cook Inlet, Alaska, for example, ice is borne on tidal currents reaching 8 knots velocity.

Ice of the North Polar icecap varies in thickness from about 9 feet in summer to 13 feet in winter, except for occasional pressure ridges which may extend as much as 90-100 feet below the surface. Throughout are small open water lakes, or "polynas" which in winter may freeze over to six inches thickness in 24 hours. The Atlantic entrance to the Arctic Ocean, east of Greenland, is broad and deep while the Bering Strait entrance from the Pacific is narrow and less than 150 feet in depth. During the early summer, pressure ridges of ice may all but block access.

THE DEEP OCEAN ENVIRONMENT

Bottom Characteristics

As on land, the topographical features of the ocean
bottom are of importance for navigational reference, the location of operational bases, their geologic implications and in
the strategy and tactics of undersea warfare.

The Continental Shelf, before dropping abruptly to
the ocean depths, may slope gently out to sea for many miles
as a submerged continuation of the Coastal Plain. Yet it is
only in the last few decades that we have learned the deep-
ocean floor has tremendous "sea-mounts", "abyssal plains",
oceanic "banks", "plateaus" and "canyons", and "ridges",
"rifts" and "fracture zones" and thus have developed a more
comprehensive picture. The estimated depth distribution of
the earth's oceans, down to a maximum of about 37,000 feet is
shown on the following figure.

Major rivers have often carved canyons in the
Continental Shelf and at the base of the continental slope
have built up great "cones" or "fans" of outwash sediments.

The principle feature of the Atlantic floor is the
Mid-Atlantic Ridge; rising as much as 11,000 feet above the
abyssal plain and breaking the surface in a few isolated
places. A deep rift valley follows its ridge and numerous
large fractures cross it normally.

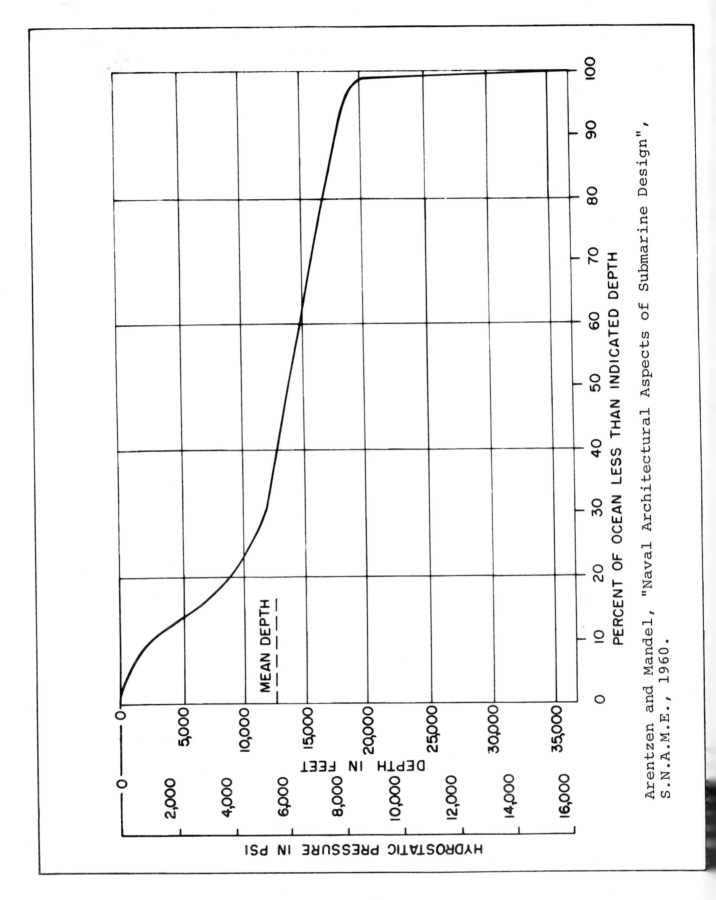

PERCENT OF OCEAN LESS THAN INDICATED DEPTH

Arentzen and Mandel, "Naval Architectural Aspects of Submarine Design", S.N.A.M.E., 1960.

The submarine barrier bisecting the abyssal plain of the Arctic Ocean, and now called the Lomonasov Ridge, was discovered in 1948-49. It effectively subdivides the waters into two regions of significantly different characteristics.

Relatively unconsolidated deposits of recent age are to be found along the shoreline of shelves off deltas. Muddy sediments of silts and clays form an inshore topstratum which grades downward into a substratum of sands and gravels which is itself laid bare along the outer portions of the shelves. Low bearing capacities (often less than 1/2 psi) and high compressibilities of the sediments are particularly striking compared to dry land values. Off the coast of Louisiana, the topstratum deposits vary in thickness to a maximum of approximately 600 feet. Warping of the strata by the growth of salt domes seriously affects the ability to extrapolate from data taken from nearby sites. Changes of elevation of a given stratum by as much as 100 feet in a distance of half a mile have been found.

Off the low sandy coasts of the North Sea the sediments commonly consist of sand that is carried toward the shore by surf and currents. This serves to build up the strand and dunes and submerged, shallow, shifting banks. Great changes can occur during severe storms.

"Slumping" is any submarine failure within the surficial sediment layers which involves movement of the material along the bottom as opposed to movement of material in suspension. Evidence is accummulating that slumping, as well as turbidity currents, constitute a major process of deep water sediment transport. Slumping has also proven to be a hazard with bottom supported structures.

When upper sediments slump they frequently expose more consolidated clays which produce abrupt changes in bottom characteristics. The irregular bottom profile produced this way can result in high axial, flexural and shear stresses in pipe lines.

Yet another problem is illustrated by the conditions in Cook Inlet, Alaska. When the powder in glacial streams makes contact with salt water, chemical reaction produces silt so fine that it largely remains in suspension as tidal currents sweep it back and forth through the Inlet. The silt makes underwater visibility nil.

Such silt on the bottom of the Inlet has very little shear strength so it is most unsuitable for supporting a drilling platform. The solution is to drive pilings 100 feet or more through the silt into the heavy clay layer beneath.

THE DEEP OCEAN ENVIRONMENT

Ocean Resources

Whereas the exploration and exploitation of outer space has been exclusively the doing of national governments, that of the ocean deeps has been overwhelmingly the work of industry; the petroleum industry in particular. World offshore oil production has already reached 16% of the total daily world output and, at present rates of increase, by 1977 it is expected to be 33% of the total output despite a threefold increase in consumption. Oil is now being produced in 340 feet of water. Fixed platforms for water depths as great as 600 feet are being devised, and floating platforms have even greater depth capabilities. Exploration is being started in depths of 1,200 to 1,500 feet. Motivation stems from the fact that the Continental Shelves, down to the 1,000 foot contour, contain 1/3 the basin area underlain by moderate to large thicknesses of possible petroleum-bearing sediments relative to similar formations for the total world land area. While total costs for offshore operations are higher than those on land, marine sediments are the more prolific.

For other mineral resources the immediate outlook is more speculative, although in the case of sulfur it is being recovered industrially at present from salt domes in the Gulf of Mexico.

Subsea underground mining already may be feasible, but for the present it appears to be uneconomical except perhaps along the coastal fringe accessible to an inclined shaft or other kind of entry from the land. Subsurface rock deposits of coal, phosphorite, oil shale, gypsum and others probably have no prospective value on the U. S. shelves for they are low value deposits for which there are other large low-cost sources. Except for magnetic ore bodies, exploration and evaluation technology is not well enough developed to discover and efficiently appraise such deposits now.

Salt in Gulf Coast domes is known to be present in large amounts recoverable at low cost. But because of large reserves available on land, it has no prospective value except in local offshore use in processing sulfur.

The surficial sediments overlaying the ocean bottom hold somewhat more promise and marine beach deposits will probably be the primary objectives of marine mining in the near future. Submarine beaches are analagous to present day land beaches and are present on the Continental Shelves because sea level during the last glacial phase, 20,000 to 40,000 years ago, was about 400 feet lower than it is today. Thus, much of the Continental Shelf was dry land and the same processes of erosion and deposition were taking place then, as now. The constant motion of beach sands provides many opportunities for the separation of sand grains of different densities. The heavy minerals are collected in large concentrations because of their greater density and are not so readily transported as the lighter quartz, feldspar and clays. Included in these "placers" are commercially important minerals such as columbium, chromium, platinum, tantalum, tin, gold, iron, silver, zirconium, diamonds and a number of rare earths. Promising submarine placer deposits occur in many parts of the world and some are being worked.

Less glamorous than the placers, but of more certain value in the near future, are shelf deposits of oyster shell, lime mud, sand and gravel. Heavy construction and glass manufacture are exhausting land supplies of sand and gravel. A recent U.S.G.S. survey shows that sand suitable for construction is present throughout most of the length of the Continental Shelf off the Atlantic Coast lying in water depths ranging from 60 to 450 feet. The thickness reaches 200 feet locally. Gravel is much less abundant, but it should be plentiful, for example, in a large fan off the New Jersey coast and in the northern margin of Georges Bank, southeast of Boston.

The well-known presence of potato-like manganese modules resting on the ocean bottom was first discovered in 1870. They have now been reported from widely scattered locations in the Atlantic, Pacific and Indian Oceans. While also rich in iron, nickel, copper and cobalt their recovery and reduction is still economically marginal.

Also among the surficial sediments are deposits of phosphorite, the major industrial source of phosphorous. Because it is necessary for the life cycle of all living things, phosphorous is an important fertilizer. Although there is no worldwide shortage, considerable efforts are being made to locate and develop the offshore phosphorite deposits found off the coasts of many countries. The explanation is that the most important factor in fixing the price of this low cost commodity is its location. Transportation costs about double its price in many parts of the world.

Adding to the uncertainties of operations in the deep ocean will be the question of sovereignty. International conventions have done much to clarify rights on Continental Shelves but for their definition, for purposes of exploitation, the Geneva Convention of 1958 prescribes, "the seabed and sub-soil of the submarine areas adjacent to the coast .... to a depth of 200 meters (656 feet) or, beyond that limit, to where the depth of the superadjacent waters admits of the exploitation of the natural resources of the said areas." With no limit set on the distance from shore or depth of sea, there is here conceivably an unfortunate invitation to get there first, reinforced by each new discovery such as the recently announced evidence of oil bearing deposits in the Gulf of Mexico at a depth of almost 12,000 feet.

Behind all estimates of future profitability are many other diverse and intangible factors. But especially in the face of the predicted exponential population growth, the conversion of many a reclamation venture from marginal to sound may be less a matter of debate and more a matter of time.

The critical shortage of copper is an illustration. Since the beginning of the century the cutoff grade for copper ores has been reduced progressively by a factor of 10; and over the history of mining, by about 250!

The sea is "water" only in the sense that water is the dominant substance present. In the following table are the approximate amounts of minerals in one cubic mile of sea water. Despite the large quantities involved in the first five items, all except sodium chloride exist in dilute amounts compared with the medium in which they are suspended. Nevertheless, it has been possible and worthwhile to extract such products as sodium chloride, magnesium and bromine from the sea.

A well-supported case has been made for the proposition that what the world needs most, that the ocean is capable of providing, is animal protein. Two-thirds of the world's population lives in countries where protein malnutrition is endemic. The bulk of the world's primary food production is in the oceans in the form of one-celled plants of microscopic size. Despite the vast volume of this "phyto-plankton", there is no practical way it can be converted directly into food. However, it forms the "grass of the ocean" upon which graze a multitude of all kinds of animals. Most of these are also microscopic in size and impractical of commercial capture and use. Upon these plant feeders live carnivores and upon the smaller carnivores live larger and larger carnivores. Each one of these stages of life is called a "trophic level" and the production ability of each level is about 1/10 that of the level beneath it in the scale. So the fish and invertebrates taken from the oceans represent about 2% of such animals actually produced. The other 98% die, decay and return to the "web of life" in the ocean, unused by man, except that in so doing their remains blanket the bottom of the oceans with commercially valuable "diatomaceous earth." If distribution were timely and equable to all people in the world, the protein supply would be more than adequate for all.

An alternative to harvesting one of the lower levels of the trophic hierarchy is the development of some form of "aquaculture" or "fish farming", by which is meant the cultivation and management of the ocean's resources in much the same way as farmers and ranchers husband the land's resources.

In 7000 years of adventuring on the seas, man too has left his mark. Off the coasts of Turkey, Greece and Italy the remains of many ancient ships have been found. A number have been systematically surveyed, relics recovered, and the scene reconstructed using archeological techniques. They are Roman, Greek, Byzantine and Bronze Age Phoencian. It has been estimated that there are at least 10,000 such wrecks in the Mediterranean and perhaps ten times as many; enough to support the expectation that one will be found to shed new light on every decade of antiquity.

As for modern times, the study and recovery of wreckage in way of sea lanes more recently heavily traveled may provide more than a picture of the culture and technology

of its time. Assuming accurate records of the ship's loss,
more precise dating can be established for similar relics
found ashore. The western reefs of Bermuda afford such
opportunities, where there are the remains of ships lost
over a period of almost four centuries.

Not only ships but large sections of cities lost
during earthquakes have been found submerged off Greece and
Jamaica.

Approximate Amounts of Minerals in One Cubic Mile of Sea Water

Sodium Chloride..................... 128,000,000 tons

Magnesium Chloride.................. 17,900,000 tons

Magnesium Sulphate................. 7,800,000 tons

Calcium Sulphate................... 5,900,000 tons

Potassium Sulphate................. 4,000,000 tons

Calcium Carbonate.................. 578,832 tons

Magnesium Bromide.................. 350,000 tons

Bromine........................... 300,000 tons

Strontium......................... 60,000 tons

Boron............................. 21,000 tons

Fluorine.......................... 6,400 tons

Barium............................ 900 tons

Iodine............................1000 to 1200 tons

Arsenic........................... 50 to 350 tons

Rubidium.......................... 200 tons

Silver............................ up to 45 tons

Copper, Lead, Manganese, Zinc...... 10 to 30 tons

Gold.............................. up to 25 tons

Uranium........................... 7 tons

From "Treasure from the Sea" by J. W. Chansler.
Included in "Science and the Sea", U. S. Naval
Oceanographic Office, 1967.

SEAWAY STRUCTURAL LOADINGS

Wave Prediction

Not much had been done in the field of wave
prediction until the Second World War when, due to the
dependence of amphibious operations upon wave conditions,
considerable effort went into the development of wave forecasting. As a result, the first of the two basic methods now
in use was developed by Sverdrup and Munk (1947). This method
is often called the "Significant Wave" method.

The wave characteristics produced are extremely
complicated and it is customary to represent them by statistical
parameters. Different statistical measures of the height distribution may be used such as "most probable", "mean", and
"significant". The "significant" wave height is defined as the
average wave height of the highest third of all the waves in
the observed sample. This particular measure was chosen and
is in general use owing to the fact that visual observations
tend to average around this value.

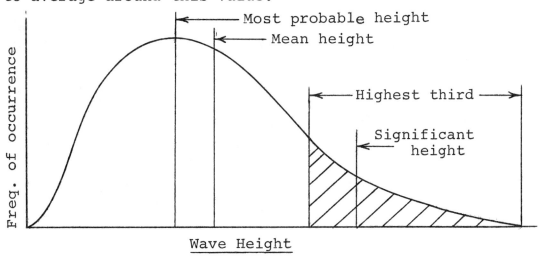

Sverdrup and Munk related the wind characteristics to wave characteristics through non-dimensional parameters of theoretically derived equations using empirical data and calculating the wave equations by determining the combined effect of the tangential and normal stress of the wind.

The relationships between non-dimensional parameters for wave height, wave speed, fetch and wind duration as presented by Sverdrup and Munk have been revised by Bretschneider in the light of additional data. To apply these relationships to a given situation, it is necessary to determine whether the situation is "fetch limited" or "duration limited." These two expressions refer to the development of a "steady state" condition for a given fetch. For every fetch, there is a minimum duration required to develop a "steady state", or maximum wave generation state, for that fetch. If the given duration is less than the duration required for the "steady state", a correspondingly lower significant wave height results. On the other hand, if the given duration is greater than the duration producing a "steady state" for that fetch, the significant wave height cannot be higher than the "steady state" height and is determined by the fetch.

The values of wave height and period obtained by the use of the fetch and duration relationships pertain to the generating area itself. The forecaster must now relate these waves to the object area for which he is forecasting, and this area can be at a great distance from the generating area.

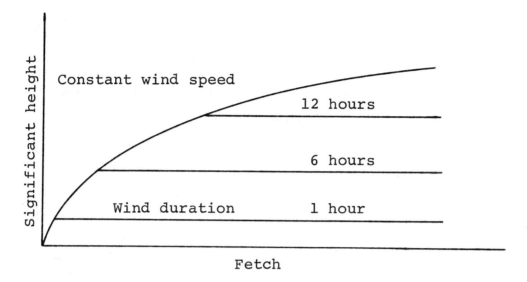

Figure representing the growth of waves and how this growth is related to the fetch and the duration.

The decay of wave height and modification of significant period which takes place between the generating area and the object area is a very complicated process. Some of the more important factors of this process are:

1. Dispersion. The spreading out due to the fact that waves of different periods travel at different velocities.

2. Angular Spreading. Waves radiate outward in different directions as they leave the generating area.

3. Wind Effects. Any winds encountered will modify the waves.

4. Encounters with other Wave Trains. There seem to be certain modifications caused by these encounters.

5. Overrunning of Currents. Currents will modify wave height.

Sverdrup and Munk's relationships for wave decay were based on theoretical considerations of the wind resistance due to the waves traveling through a region of calm. Bretschneider revised these relationships in 1952 from an empirical point-of-view including in the relationships an additional factor, the minimum fetch. By minimum fetch, one means; that fetch which would produce the same wave height in a duration limited case, or the actual fetch in a fetch limited case.

The spectral method is based on an energy spectra. This method is also known as the Pierson-Neumann-James method. This approach describes the sea surface as a result of the combining of an infinite number of sine waves of various amplitudes, frequencies, and directions. The wave characteristics formed are a summation of these sine waves and are described by a Gaussian function. The energy, E, of each spectral frequency is equal to the square of the amplitude, A, of the particular wave train. The frequency band of maximum energy is dependent upon the velocity of the wind in a fully arisen sea ("steady state").

The process of wave growth in the generating area has been related to the co-cumulative spectra which is the integral of the energy spectral curve starting from the high frequency end.

The limitations on wave build-up imposed by fetch or by duration are represented by constraints on the low

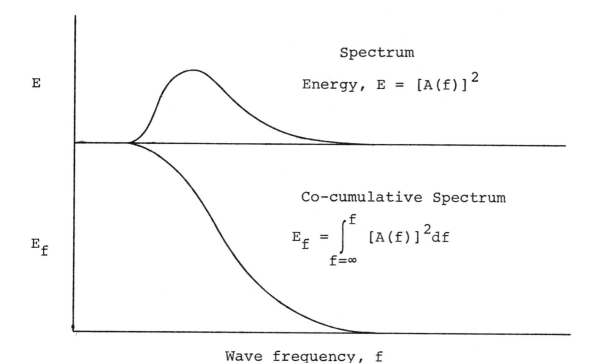

Wave frequency, f

frequency range of the spectra. That is to say, wave growth
starts with high frequency components and works towards the
low frequency components. The total energy, which is calculated from the co-cumulative spectra, is the basic statistical
quantity used in forecasting and it is related to the other
quantities by the following simple formulae:

Significant wave height $H_s = 2.83 \sqrt{E_f}$

Average (mean) wave height $H_{ave} = 1.77 \sqrt{E_f}$

1/10th highest wave $H_{1/10} = 3.60 \sqrt{E_f}$

where heights are in feet and E_f is in $(feet)^2$.

It is in the propagation of waves from the generating
area to the object area that the spectral method offers the
greatest advantage.

In the real ocean, wind speeds are not constant by any means, and generating areas are not static. The spectral method has proposed a series of "filters" to handle these problems. In each case, the filter represents the selection of frequencies in the wave spectra which corresponds to the particular situation.

The procedure is to select the frequencies that are to be transmitted; transmit them from the generating area to the object area taking into account decay by dispersion and angular spreading; and then reconstruct a forecasted spectra at the object area.

Propagation of waves into shallow water will modify the spectra and individual form of the waves.

Refraction will modify the spectra of waves just as it does light. As the waves "feel bottom" energy losses will occur and for an individual wave, linear theory is no longer useful for predicting wave amplitudes.

In shallow water a sinusoid is representative of a real wave only in period. There are many higher order wave theories which can be applied.

SEAWAY STRUCTURAL LOADINGS

Wind, Wave and Gravity Forces

Although structural loadings (other than hydrostatic pressure) can not be estimated by any such simple means, they are nevertheless amenable to individual determination and superposition because any relationship between them is, at most, indirect. It falls to the designer, then, to establish the proper magnitude of each for each design condition and use the worst combination whenever and wherever in the structure it is critical. Rather than presenting the rigorous but extended development of the subject, this introductory treatment deals with it in the more general terms of "design loads" useful in preliminary stages of structural design.

Wind Loads

The wind force, F, in pounds, can be found from

$$F = A\frac{wv^2}{2g}$$

where

A = effective projected area, ft.^{2}

w = unit weight of air
(0.0765 lbs./ft^3 at sea level at 59°F)
(0.0807 lbs./ft^3 at sea level at 32°F)

v = wind velocity, ft/sec.

g = gravity acceleration
(32.2 ft./sec.^{2})

The force produced on a given projected area, A_p, is a function of the flow patterns produced and hence of the aspect ratio, the wind velocity and the configuration and orientation of the surface presented. The relationship between the actual and effective projected areas is usually expressed as

$$A = C_s A_p$$

in which C_s is a drag or shape coefficient empirically derived from model experiments. A few illustrative examples are given in the following table. For elongated bodies, such as ships, and intermediate angles of attack, for example, specific model tests may be necessary, especially for swinging moments and, of course, for multiples of units which are closely spaced.

Wind velocities as a function of elevation above the water are given by the following, where H_1 and H_2 are the two elevations concerned.

$$v_2 = v_1 \left(\frac{H_2}{H_1}\right)^{\frac{1}{7}}$$

For purposes of hydrostatic stability, overturning moments at a number of angles of heel and all critical operating conditions must be found.

Current Loads

The expression for the wind force is equally suitable for currents, after due allowance is made for the unit weight of sea water at 64 lbs/ft^3. However, model tests show that once the

TABLE 1

Values of Shape Coefficient, C_s

	Length/Width Ratio		
	1	5	∞
Flat plate (perpendicular to flow)	1.16	1.20	1.90
Prism (axis perpendicular to flow)			
Square			2.00
2:1 (long side parallel to flow)			1.50
Cylinder (axis perpendicular to flow)			
Reynolds No. = 10^5	0.63	0.74	1.20
Reynolds No. = 5×10^5		0.35	0.33

American Bureau of Shipping Design Values

Cylindrical Shapes	0.50
Surface ship hulls, deck house sides, underdeck areas (smooth surfaces)	1.00
Rig derrick (each face)	1.25
Underdeck areas (exposed beams and girders)	1.30
Isolated structural shapes (cranes, angles, channels, beams, etc.)	1.50

water depth is less than six times the vessel's draft, the shape coefficient will rise exponentially to magnitudes greater by a factor of about six when the depth/draft ratio is one.

Wave Loads

The classical theory for wave forces on cylindrical bodies sees the total force as being composed of a drag component due to shear and an inertial component due to fluid acceleration. Thus, for a small cross sectional element of axial length, ds, in a circular cylinder of radius D, the total load is:

$$dF_T = dF_D + dF_I$$

where

$$dF_D = C_D \frac{w}{g} D \frac{u^2}{2} ds$$

and

$$dF_I = C_M \frac{w}{g} \frac{\pi D^2}{4} \dot{u} ds$$

in which

C_D = "drag" coefficient

C_M = "inertia" coefficient

u = horizontal velocity of water particles

$\dot{u}$ = horizontal acceleration of water particles

w = unit weight of sea water (64 lbs/ft^3)

Values of C_D and C_M for simple shapes have been experimentally derived. Typical values of C_D may be seen in the table of values for C_S given above.

The American Bureau of Shipping quotes typical values for C_M as given in the next table.

Particle velocities and accelerations as a function of depth can only be predicted from one of the several available wave theories and there are none which satisfy the dynamic free surface boundary conditions throughout the range of offshore sea conditions which may be anticipated. The relationship primarily governing suitability is the ratio of water depth (in feet) to wave period (in seconds) squared, with shallow, intermediate and deep water waves each falling into a different category.

Even if the determination of particle velocities and accelerations at a sequence of depths were not by themselves laborious, the integration over the lengths of all members to arrive at total forces and overturning moments, as is necessary ultimately, would make automatic computation almost a necessity. Furthermore, it is usually necessary to investigate for several wave approach angles, wave periods and crest locations.

From the nature of the distributions of these velocities and accelerations with depth, it is evident that the wave force on a plane, ocean floor support mat, for example, is nearly all inertial.

Vertical force differentials with the passage of waves may also be significant. The principal factors are the lift from inclined members and buoyancy changes.

TABLE 2

Values of Inertia Coefficient, C_M

The tabulated values below are for two dimensional flow. Values for three dimensional flow may be approximated by multiplying by the correction factor, K.

With elliptical and rectangular shapes, h is the dimension parallel to the flow, b is the breadth normal to the flow and l is the length normal to the flow. The "lb" plane is the plane normal to the flow.

Cross Sectional Shape	On the Surface	Submerged	On the Bottom
Flat plate (with cross sectional area = $\pi b^2/4$)		1.0	
Circle			
Small Diameter		1.5	
Large Diameter		2.0	
Ellipse		1.0+b/h	
Rectangle			
With l horizontal	1.0+b/2h	1.0+b/h	1.0+2/h
With l vertical	1.0+b/2h	1.0+b/h	

$$\text{Correction factor, } K = \frac{(l/b)^2}{1+(l/b)^2}$$

Ice Loads

As a result of investigations in Cook Inlet, Alaska by
the major oil companies, an average crushing pressure for ice
was found to be about 200 lbs/in^2. Using an ice thickness
of 3.5 feet, this results in a design loading on the leading
legs of a structure amounting to 100,800 pounds per foot of leg
diameter. Peyton indicates that ice strength is widely variable
and, among other things, is rate dependent. Crushing strength
can be considerably higher than quoted above and the force so
generated may oscillate at one cycle per second.

It was also felt necessary to anticipate an impact force
from a large iceberg borne by the currents. According to
Daigle, "this force was derived by equating the kinetic energy
of an ice mass to the work required to cause an indentation in
the ice mass by the tower leg." For this purpose a block 50
feet square and 20 feet thick was assumed; traveling at a speed
of 9 feet/sec.

Another important aspect of ice loading is the formation
of masses encircling the legs which, upon thawing, may ride up
and down with the waves and tides until becoming hung up on a
joint or other surface discontinuity. Sizable impact loadings
and an added component of weight are the result.

Gravity Loads

Simply as order-of-magnitude deck loadings, the American
Bureau of Shipping prescribes the following as minimum values

TABLE 3

	Design Loadings (lbs/ft^2)	Equivalent Head of Sea Water (ft.)
Crew spaces (walkways, general traffic areas, etc.)	90	2
Work areas	180	4
Storage areas	270	6
Helicopter platform*	40	0.9

*Plus 75% of helicopter gross weight on each of two points one square foot in area or the manufacturers recommended wheel loading.

SEAWAY STRUCTURAL LOADINGS

Settlement, Scour and Earthquakes

The term "Foundation Engineering" has come into use to encompass that branch of engineering concerned with the evaluation of a potential building site and the incorporation of the conclusions into the foundation design. Of primary concern in the ocean environment are subsidence, settlement (especially of differential kind) and scour.

Site investigations can be initiated by a study of available geological and oceanographic reports which may include bottom and sub-bottom soundings and the geological history of the area. On site studies rely principally upon bottom coring, bottom surface sampling and subsequent laboratory evaluations.

Soil Properties

Soil properties are classified either as engineering or index properties. Engineering properties of interest are "permeability"; the resistance to flow of water through the void spaces, "Compressibility"; the resistance to change of volume under load, and "Deformability and Shear Strength"; the resistance to change in shape and rupture, with or without volume change, under applied load. Index properties, which may be measured more cheaply and quickly, are commonly used for making qualitative estimates of the engineering properties.

Categorization of soils as "coarse-grained," fine-grained,"
and "highly organic" is supplemented by quantitative index
values of water content, void ratio, porosity, grain-size
distribution and Atterberg limits (which relate to water
contents at various arbitrarily defined consistencies of
clay). With the soil properties known, a number of
theoretical and emperical solutions exist for predicting
soil behavior in terms of the stress distribution under a
bearing element, the steady-state flow of water through
soil, the consolidation of clay strata and the bearing
capacities of footings and piles.

In the case of the ocean floor these overall
procedures are less secure; first, because the topography
and geology are less well known in detail and only the
broadest of generalizations can be made. The practical
difficulties of sampling for considerable depths below the
ocean floor foundation level are obvious and a further
complication is obtaining undisturbed samples of any kind
no matter what the location. In addition, on-the-bottom
conditions, especially as regards water content, are almost
impossibly difficult to duplicate in laboratory tests,
although given the same material ingrèdients. It is noteworthy that, in general, the higher the water content the
lower the soil strength.

DESIGN OF OFFSHORE MOBILE PLATFORMS

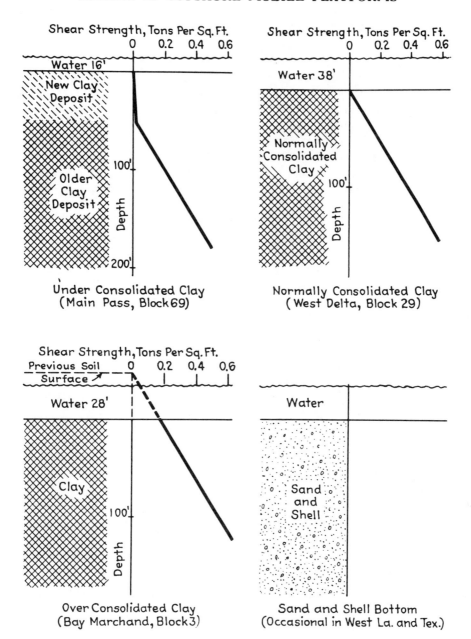

Typical Soil Profiles for the Gulf of Mexico
(Rectin, Steele and Scales, S.N.A.M.E., 1957)

Foundation Support

For the sake of defining concepts, the vertical support force exerted on a pile for example, is composed of the shearing resistance of the soil along its sides and the bearing pressure of the soil under its tip. In clay, the skin friction component, which is a function of the exposed area and the shear strength of the soil, overwhelmingly predominates. In sand, quite the opposite is apt to be true, stemming in part from the fact that the required pile penetrations in sand are usually small. Layering of materials on the ocean floor necessitates the specific profile being known for any reasonably rational evaluation of bearing load capacity.

When large bearing areas are involved, as with a mat structure, a significant additional component of support due to "buoyancy" comes into effect with any penetration of the mat into the soils.

Failure in a cohesive soil, such as clay, under a vertical bearing load is presumed to occur in shear along an interface as shown below.

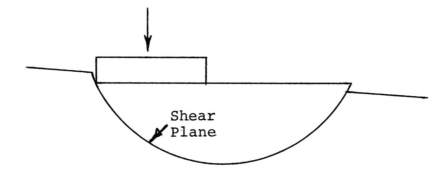

Shear
Plane

If the footing was to be infinitely long, the unit bearing load for failure is on the order of 5.7 times the average shear strength of the soil. If the footing is square, the value is 7.4 owing to the shear also taking place at the ends of the foundation.

Resistance of a bottom supported mat to lateral displacement along the floor is, first of all, due to the shearing strength of the soil along the bearing surface and is directly related to the bearing area. This resistance may be augmented through the addition of piles or spuds driven through the mat into the supporting soil. Rechtin, et al, give a relationship for the resistance afforded as,

$$R = 11 \ LDQ_c$$

which is merely the product of the length of pile penetration, the pile diameter, the soil shear strength from triaxial laboratory tests and an emperically derived factor. As the pile penetration is increased and pile flexibility becomes greater, its lower portion is less effective. The simple expression above is then quite inadequate and will over-estimate the resistance.

As a practical design condition, the ABS offshore platform rules specify that legs of self-elevating units, if they have no individual bottom pads so that they penetrate the sea bed, for structural analysis must be assumed to have a pinned end condition at a point 10 feet below the ocean floor.

Scour

Scouring erosion is of concern for its potential
to undermine a mat foundation, for example, or to lower
the effective fixation point in piles. Some of the factors
involved are listed by Rechtin, et al, as depth of water,
mat propertions relative to wave size and water depth,
whether due to current or wave action, water velocity and
the soil properties. A conventionalized system has been
adopted whereby various sizes of boulders and cobbles and
particles of gravel, sand, silt and clay have been assigned
the fall velocity in water of a quartz sphere of equal
diameter. While no direct comparison can be made between
fall velocity and scour velocity, it is noted that the velocity along the bottom necessary to cause general bed movement is roughly twice the fall velocity.

To allow for the possibility of scouring, the
ABS platform rules state that in the design of a single
footing, of the broad mat type, 20% of the area is to be
considered unsupported by bottom bearing. When there are
two footings or more, any one footing is to be ocnsidered
unsupported on 50% of its area.

Breakout Forces

Owing to bottom suction and silting-in, the
force necessary to remove the mat of a mobile drilling rig,
for example, may be many times the weight of the object.
Little of a quantitative nature can be said about the

evaluation of their magnitudes except that they are rate
dependent and may be drastically reduced with lengthened
removal time.

Earthquakes

In earthquake prone areas such as the west coast
of the United States, wind and wave loadings may even be of
secondary importance. Attempts have been made to define
rational structural design criteria for structures in this
area and they usually yield force magnitudes in "g's"
applied at the base of the structure as a function of
occurrence probability. Some designers choose some
modification of the El Centro (1940) accelerogram as the
design earthquake, perhaps after attempting to account
for the actual location of the site in question relative to
a likely hypocenter, the variability of soil characteristics,
and trying to introduce some consistency with wind and wave
forces by basing all on the same periodicity of occurrence
probability, say a 100 year likelihood. Although the
El Centro data represent the largest strong motion record ever
obtained, questions can be raised as to its general applica-
ability and how such modifications to it can be effected.
As an alternative, a generalized strong motion earthquake
spectrum has been proposed by Wiggins which may be used
with whatever occurrence probability, site conditions, earthquake magnitude and hypocentral distance are pertinent.

SEAWAY STRUCTURAL LOADINGS

Longitudinal Strength

Full scale experiments with ships of many cross
sectional configurations have verified that under longitudinal bending the stress distributions follow simple
beam patterns to a remarkable degree; regardless of the type
of framing system or any of the other characteristics likely
to be variables. For the length/depth proportions of the
vehicles involved, there is little doubt but that, of the
component stresses, bending stresses rank well above vertical shearing stresses in importance.

With surface craft, longitudinal strength is a
vital consideration and the most taxing condition occurs when
traversing a wave whose length is about the same as that of
the vehicle. With the elongated, spar-like platforms now
finding use as stable research stations, another operating
condition may be equally significant. After being towed in
horizontal attitude to a work site, it is customary for the
platform to be reoriented vertically by ballasting. The
manner in which this is done has structural as well as
hydrostatic implications.

To an even greater degree than with ships, it is
the worst reasonable single load application which will be
structurally governing; rather than some loading spectrum
leading to fatigue or other form of progressive failure.

This seems to be a reasonable conjecture judging from the very small percentage of its life that a spar platform will be other than upright and there subjected to very little bending at all.

The up-ending operation is unlikely to take place except in calm conditions and so is essentially one in which there is hydrostatic equilibrium at each stage of the procedure. Until very recently it was also usual to estimate shearing forces and bending moments for in-transit, wave conditions based on a static equivalent wave sufficient to include such dynamic effects as the rigid body motion response, orbital motions of water particles, random sea effects, interaction between the floating body and the water, etc. Two worst conditions; on the crest of the wave (hogging) and in the trough of the wave (sagging) are to be anticipated.

High frequency stress variations, although secondary, may be significant augments, but the effect of the elastic deflection of the "rigid" body on primary bending is of no consequence.

Given the immersed shape distribution for equilibrium (and hence the buoyancy distribution) and the longitudinal weight distribution, the shearing forces and bending moments at all points may be found in the usual way, viz.

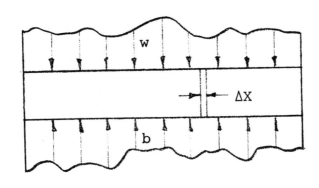
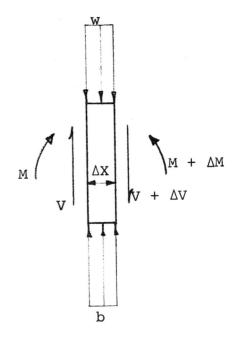

$$\sum F_V = 0 = \cancel{V} - (\cancel{V} + \Delta V) + (b-w)\Delta X$$

$$\Delta V = (b-w)\Delta X$$

or $\quad \dfrac{\Delta V}{\Delta X} = (b-w)$

and $\quad V_X = \displaystyle\int_0^X (b-w)\,dX \; \cancel{(+ C_1)}$

Since at X=0, $\quad V_X = 0, \quad \therefore \quad C_1 = 0$

Also

$$\sum M = 0 = \cancel{M} - (\cancel{M} + \Delta M) + (V + \cancel{\Delta V})\Delta X - (b-w)\frac{\Delta X^2}{2}$$

$$- \Delta M + V \Delta X = 0$$

$$\frac{\Delta M}{\Delta X} = V$$

and $\quad M_X = \displaystyle\int_0^X V\,dX \; \cancel{(+ C_2)} = \int_0^X \int_0^X (b-w)\,dX\,dX$

Since at X = 0, M = 0 $\quad \therefore \quad C_2 = 0$

- 46 -

Except for such simple forms and weight distributions as in the ballasted cylinder of the illustration, integration by numerical or graphical methods may be required. (The concentrated weight of the working platform and equipment is represented by the force "W".)

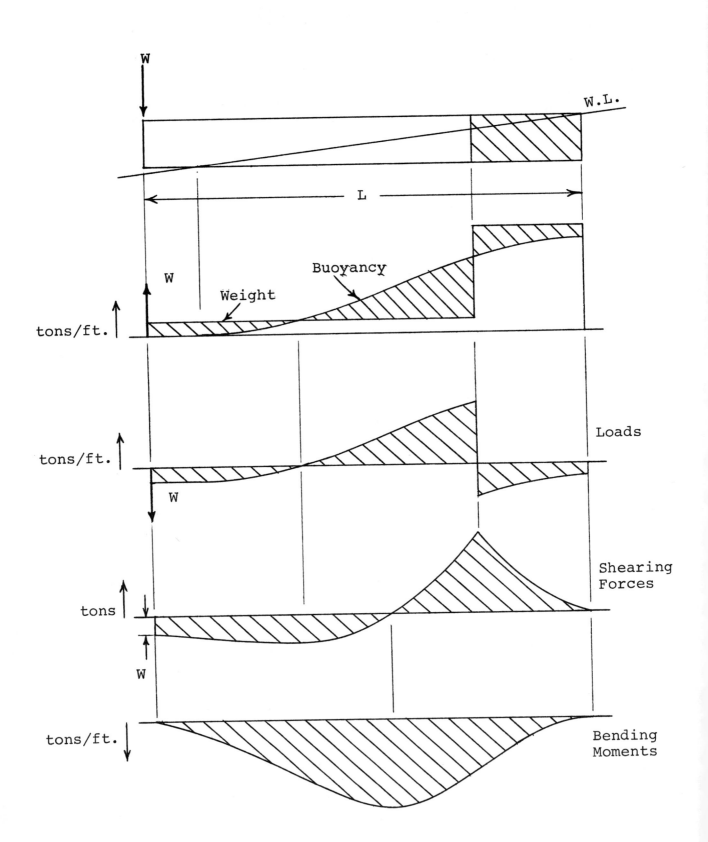

SEAWAY STRUCTURAL LOADINGS

Rigid Body Oscillation Forces

The motion of a floating body in a seaway gives rise not only to virtual changes in the weight of concentrated masses on board, but also to force components in directions other than perpendicular to the vessel's base line plane. Wracking forces and foundation design will require a detailed appreciation of these oscillation induced loadings.

Axes will be taken fixed in the floating body with the origin at the center of oscillation. Of the angular oscillations, "roll" is about the fore and aft axis OX, "pitch" is about the transverse axis OY, and "yaw" is about the vertical axis OZ.

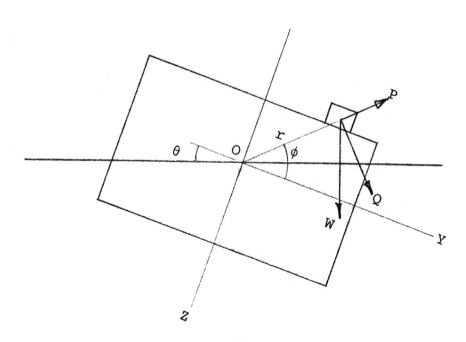

In simple roll, the centrifugal force, P, the tangential force, Q, and the weight, W, can be resolved merely into transverse and vertical components. Total forces in these and in the fore and aft direction are the most useful form of the information needed.

Thus,

$$P = m \dot{\theta}^2 r$$

and the components are

$$P_y = m \dot{\theta}^2 r \cos \phi = m \dot{\theta}^2 t$$

and

$$P_z = m \dot{\theta}^2 r \sin \phi = m \dot{\theta}^2 v$$

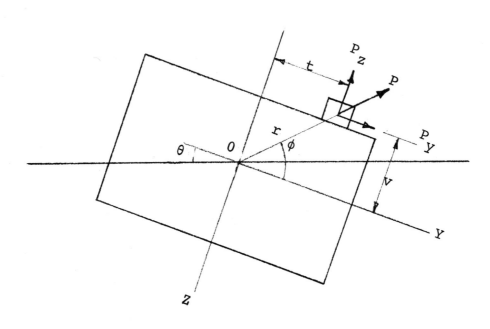

Similarly,

$$Q = m \ddot{\theta} r$$

and the components are

$$Q_y = m \ddot{\theta} r \sin \phi = m \ddot{\theta} v$$

and

$$Q_z = - m \ddot{\theta} r \cos \phi = -m \ddot{\theta} t$$

for the case illustrated. (With positive linear distances being assumed below and/or to the right of the origin, forces are also positive when in the direction of increasingly positive linear distances.)

Hence, in the general case, for rolling alone, the total components are

$$R_y = \pm \, m \, g \sin \theta \pm m \, \dot{\theta}^2 t \pm m \, \ddot{\theta} v \qquad (1)$$

and

$$R_z = + m \, g \cos \theta \pm m \, \dot{\theta}^2 v + m \, \ddot{\theta} t$$

If, in the case of simple pitching, α is the angular displacement and ℓ is the fore and aft distance from the center of oscillation to the mass, then, likewise

$$R_x = \pm \, m \, g \sin \alpha \pm m \, \dot{\alpha}^2 \ell \pm m \, \ddot{\alpha} v \qquad (2)$$

and

$$R_z = + m \, g \cos \alpha \pm m \, \dot{\alpha}^2 v \pm m \, \ddot{\alpha} \ell$$

When rolling and pitching occur together the R_z total becomes

$$R_z = + m \, g \, \cos\theta\cos\alpha \pm m \, \dot\theta^2 \, v \pm m \, \ddot\theta \, t \pm m \, \dot\alpha^2 \, v \pm m \, \ddot\alpha \, \ell$$

(3)

From the fundamental equation of motion for unresisted rolling in still water comes an expression for angular acceleration which, for small angles, may be reduced to the following

$$\ddot\theta = - \frac{\Delta\overline{GM}\theta}{I_T}$$

in which

Δ = floating body displacement

$\overline{GM}$ = floating body metacentric height

I_T = mass moment of inertia of the floating body

The general solution to the differential equation also yields the expression for the complete period of roll, viz.

$$T_R = 2\pi \sqrt{\frac{I_T}{\Delta\overline{GM}}}$$

Substituting for I_T in the acceleration expression gives

$$\ddot\theta = \frac{4\pi^2\theta}{T_R^2}$$

The relationship between angular velocity and angular acceleration being such that when one is a maximum the other is zero, it would seem that the largest, and hence the design values, of the three total components would be found at or near the angular displacements at which the above relationships occur. That is, in rolling, when

$$\theta = 0 \ , \ \dot{\theta} = \dot{\theta}_{max} = \frac{2\pi\theta_{max}}{T_R} \ (\frac{\pi}{2} \ x \ average \ \dot{\theta}) \ and \ \ddot{\theta} = 0$$

and when

$$\theta = \theta_{max}, \ \dot{\theta} = 0 \ and \ \ddot{\theta} = \ddot{\theta}_{max} = \frac{4\pi^2\theta_{max}}{T_R}$$

On the face of things, then, for the general case of simultaneous rolling and pitching, the exercise consists of two evaluations each for R_x and R_y in order to find the larger of each. Signs are chosen by inspection to suit the location of the concentrated mass, whether or not a simple half cycle or continuous oscillation is expected, i.e. generally in anticipation of the worst reasonable combination of conditions. Once a maximum angle of roll has been chosen, only the rolling and pitching periods, T_R, and T_P, are unknown and these must be estimated. With surface ships, average values for each class of vessel are well known and maximum values of roll angle for design are often taken as 30° in defiance of the small angle limitation on the solution as well as its being confined to unresisted motion in still water.

R_z forces require somewhat more attention as there are obviously four combinations of roll and pitch angles to be considered. Furthermore, should the mass be located above the center of oscillation the case of $\theta = 0$, $\alpha = 0$ may be eliminated as it yields a value of R_z less than the static weight, $+ m g$. For the same kind of reason, in the two other cases involving either $\theta = 0$ or $\alpha = 0$, the residual centrifugal force terms should be eliminated in anticipation of the possibility that rolling could take place without pitching and vice versa.

Of course, the forces due to the translational oscillations of heave (parallel to OZ), surge (parallel to OX) and sway (parallel to OY) should be added, with allowance for coupling effects, but up to the present it has not been found feasible to treat these components separately in estimating design loadings. This is because the oscillation periods, usually as measured on full scale ships by crude means, have not been sufficiently discriminating to distin- guish between the component motions involved.

With self-elevating drilling units, the requirements of the American Bureau of Shipping call for the legs, when in field transit position, to be able to withstand a "bending moment caused by a 6° single amplitude roll or pitch at the natural period of unit plus 120% of the gravity moment caused by the legs' angle of inclination." In ocean transit position they are to be "designed to withstand a bending moment which is equal to 110% of the maximum righting moment of the unit (in roll or pitch) divided by the number of legs plus 120% of the gravity moment caused by the legs' angle of inclination."

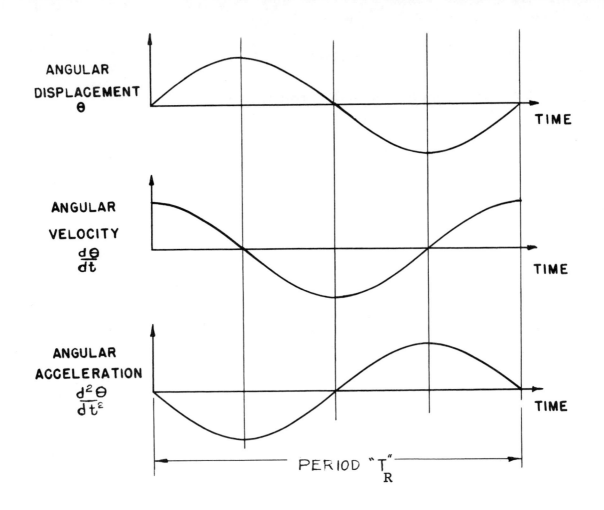

ANGULAR DISPLACEMENT θ

ANGULAR VELOCITY $\frac{d\theta}{dt}$

ANGULAR ACCELERATION $\frac{d^2\theta}{dt^2}$

TIME

PERIOD "T_R"

VARIATION OF R'_Z WITH ANGULAR DISPLACEMENT

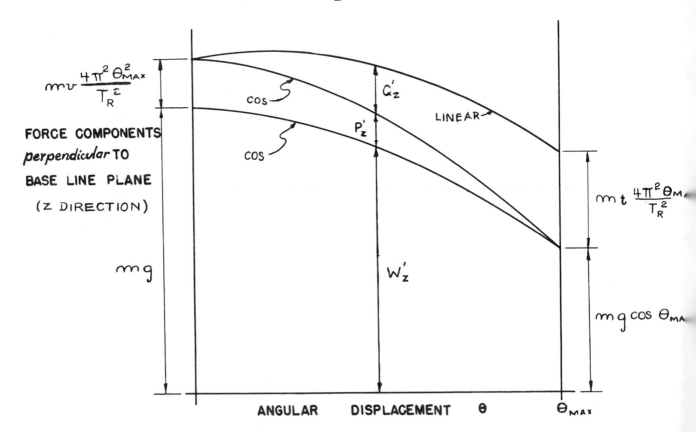

$m v \dfrac{4\pi^2 \theta^2_{MAX}}{T_R^2}$

FORCE COMPONENTS *perpendicular* TO BASE LINE PLANE (Z DIRECTION)

mg

COS

COS

G'_Z

P'_Z

LINEAR

W'_Z

$m t \dfrac{4\pi^2 \theta_M}{T_R^2}$

$mg \cos \theta_{MA}$

ANGULAR DISPLACEMENT θ

θ_{MAX}

MATERIALS

Required Properties and Promising Materials

The selection of materials for any given ocean engineering application is partly the problem of determining which of the many material characteristics are, in sum, most suitable to the specific structural problem. This process is a primary design consideration for marine structures, particularly in view of the wide variety of applications presently found in this field, the diversity of structural characteristics demanded of each, as well as the hostile environment in which they are required to exist and operate.

The weight density of a material is very often a critical characteristic since structural weight is so often a major design consideration. In many cases it is not the absolute density itself which is important but a strength to weight ratio, usually represented by the ratio of either yield stress or ultimate stress to the weight density. Such a parameter is usually employed in cases where it is desired to maintain a certain level of strength with minimum structural weight.

Both ductility and notch toughness are measures of a material's ability to absorb energy through plastic deformation, without fracturing. Notch toughness refers to the ability of a material to resist brittle fracture in the presence of metalurgical or physical cracks or notches. In general, the more energy absorbed, the more ductile or tough

the material is said to be. When little or no energy is
absorbed plastically before fracture and the break is of the
cleavage type, the material's behavior is described as "brittle".
Many materials exhibit what is termed a "transition temperature",
a temperature below which the material's behavior changes quite
rapidly from ductile to brittle, and its energy absorption
capabilities fall off equally abruptly. A material with high
transition temperature is usually described as "notch sensitive".
Since for many materials the transition temperatures are in the
same range as certain ocean temperatures, this phenomenon is of
particular concern to ocean structures.

Loads or deformations which will not cause fracture
in a single application can result in fracture when applied
repeatedly. This mechanism of fatigue failure is complex, but
basically involves the growth and propagation of small surface
cracks. It is customary to designate the stress which can be
withstood for some number of cycles as the "fatigue strength"
of the material at that point. Likewise the number of cycles
which can be sustained at a given stress level is the "fatigue
life". As the number of stress cycles experienced by most
ocean structures in their lifetime may be rather small, low
cycle fatigue is likely to be a most important part of the
spectrum. Some materials have the property of a stress level
which the material can withstand for an indefinite number of
cycles without failure. This stress level, corresponding to
essentially infinite life, is termed the "endurance limit."

Other material characteristics which merit consideration include corrosion properties of many facets, ease of
fabrication, joinability, including weld-ability, durability,
maintainability, general availability, and finally, cost. With
several possible modes of failure to be anticipated in each
element of a structure and weight and/or cost to be minimized
(or perhaps other performance characteristics to be optimized)
trade off studies must be resorted to before a final optimum
choice of material can be made for any specific application.

The most primising structural materials fall into
three main categories; metals, nonmetals, and composites.
Steels - Steels show promise mainly because of the extremely
high strengths which new heat treatment techniques are making
possible. These new steels include such types as HY80, HY100,
HY150, and the maraging steels. Yield stresses run from
80,000 psi for HY80 to approximately 400,000 psi for some
maraging steels. A tendency towards brittle behavior and low
notch toughness, in addition to only moderate enhancement of
fatigue life are the major drawbacks of these high strength
steels.

In general, metals well suited to stagnant conditions
are quickly eroded in the presence of increased velocities of
aerated sea water. Likewise, metals such as the nickel-
chromium-iron alloys perform well at high velocity, but crack
and pit in stagnant conditions.

Aluminum - Aluminum is of interest mainly because of its low density. Some of the new aluminum alloys are competitive with some steels in yield and ultimate strength, but have better corrosion resistance. As with steel, as strengths increase, aluminum alloys show the tendency towards increased brittleness, lower notch toughness, and questionable fatigue life.

Titanium - Titanium combines a relatively low density with very high strength, excellent fatigue properties and corrosion resistance, and anti-magnetic properties. A severe problem with titanium is stress-corrosion cracking. Especially for deep applications, titanium alloys are generally considered to be the most promising materials in the hydrospace field despite present high cost.

Copper-Nickel Alloys - Copper-Nickel alloys are valued in applications not requiring very great strength levels, for their excellent corrosion resistance properties, at least under stagnant conditions.

Glass and Ceramics - Glass and ceramics are of interest because of their extremely high strengths in compression. They also demonstrate excellent corrosion resistance qualities. Glass, in addition, offers the advantage of transparency. The chief drawbacks of glass and ceramics are their brittle behavior and low fracture toughness.

Composite Materials - Composite materials are made of filaments of some material specifically oriented in a matrix material. The filaments may be of either a metallic or

non-metallic material. Glass and boron are commonly
considered. Such fiber composites are being developed with
very high strength to weight ratios. As with glass and
ceramics, the problems of fastening and joining and of
delamination under pressure in long term use is currently a
major problem with composites.

Other Materials - Plywood and concrete have been suggested
for use in underwater sturctures. The main advantage of both
is their relatively low cost. In addition, concrete posseses
good compressive strength, good availability, resistance to
corrosion, and excellent formability. Its chief disadvantage
is its limited tensile strength.

LINE STRUCTURES

Anchors, Chain and Line

A logical starting point for a discussion of ocean
engineering cable systems is a description of the elements
of such systems and their individual characteristics.
Properties of anchors, chain, metallic cable, and fiber
rope will be reviewed insofar as these properties have a
profound influence on the behavior of both simple and complex
cable systems.

Anchors have existed from ancient times. The more
conventional anchors depend for their functioning upon a
submerged weight sufficient to develop a bottom friction
adequate for the holding power required. The flukes serve to
augment this gross friction coefficient. These statements
imply a relatively flat angle of load application, either
through the use of large scope in the mooring or a carefully
contoured mooring trajectory. The first anchor was probably
a rock, large enough so that its submerged weight alone was
sufficient to hold position. This and other types of "clump
anchor" are still used for small craft and in harbor moorings.
In an attempt to improve on the clump anchor's relatively low
holding power to weight ratio, the design of anchors, developed
through the years, has evolved in the addition of flukes of
varying style. The four most common types of anchor are the
stock, the stockless, the Danforth and the mushroom anchors.

PRIMITIVE STONE "ANCHOR"

KILLICK

EAST INDIAN

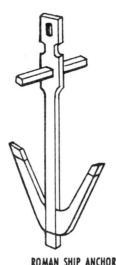

ROMAN SHIP ANCHOR

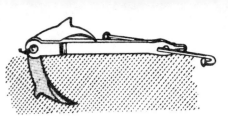

PORTER

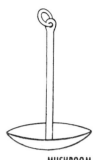

MUSHROOM

KEDGE ANCHOR

PLOW OR CQR

CHINESE

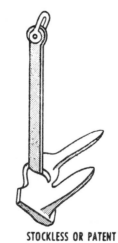

STOCKLESS OR PATENT

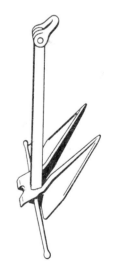

DANFORTH

ANCIENT GREEK (FROM COINS)

WISHBONE

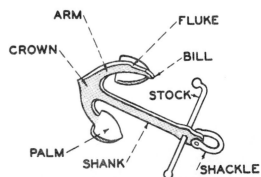

ARM   FLUKE

CROWN   BILL

STOCK

PALM   SHANK   SHACKLE

These four all have the common objective of developing
maximum holding power on a minimum anchor weight. The
presumed bottom characteristics and ease of recovery are
other design considerations.

Development of new anchor designs has by no means
stopped with the above. Between 1955-58 the U. S. Navy
developed a family of new mooring anchors called the
Budocks Stato which possess a very high relative holding
power to weight ratio. In addition, other new types of
anchors are being tested. These include: (a) drilled in
and cemented pilings; (b) driven piling; (c) explosive
anchors; and (d) the jetted in dead man. These new anchors
have an important factor in common in their favor; their
ability to withstand an upward force. Also, some of them
are a great deal more reliable. These factors obviously
imply permanence or long term use and expendability.

While considerations such as expense, ease of
handling and storage will influence the choice of an anchor,
the single most important characteristic is the anchor's
holding power. It is this fact which makes the holding
power to weight ratio so important. For any given anchor
this ratio is not fixed, but depends particularly on the
nature of the ocean bottom. Certain anchors are designed
for certain types of bottom conditions and therefore may
not work well when they encounter other than the design
type surfaces. Formulas to estimate roughly the holding

power of various types of anchors for different bottom
conditions have been developed, but for any given situation
the actual holding power of an anchor should be determined
in tests at the site,if at all possible.

Chain has traditionally been used as mooring line
for naval vessels, but suffers the handicap of a low strength-
to-weight ratio which effectively removes it from consideration as mooring line for lighter oceanographic equipment
installations. Nevertheless, it has other properties which
make it extremely useful in certain underseas applications.
One is its complete lack of bending stiffness which makes it
an excellent flexible or slope-matching connector. Another
is its inability to support any longitudinal compression.
In addition, chain is also more resistant to abrasion damage
than cable or rope and thus is often used for ground tackle.
In fact, its high mass density and the resultant catenary
formed can also be used to advantage in soft moorings where
minimizing snubbing shock is desired.

Metallic cable has until recently been used almost
exclusively for underseas lines. Wire rope, mostly of
stainless or improved plow steel, is used for most cable
systems in view of its higher strength-to-weight ratio and
its favorable handling characteristics for a given strength
level. However, metallic cable has some important drawbacks.
It is susceptible to fatigue failures due to stress corrosion
and is also vulnerable to dissimilar metal and oxygen-cell

attack. These problems are particularly troublesome in the
areas near terminal fittings and clamps and where kinks have
been formed in the cable. Since kinking occurs in twisted
cable as a result of its tendency to untwist under load,
slackening or flexing of cables is to be avoided and care
must also be taken to protect them from abrasion, particularly near ground tackle on the bottom, if premature
failures are to be avoided. Metallic cable is also subject
to severe fretting (separating of the strands) due to passage
over sheaves while under tension. While this problem may be
eased somewhat by use of large-diameter sheaves, this problem will remain until better methods of handling cable
than winching are devised.

Natural fiber rope has fallen almost completely
out of use in undersea work, since it is highly susceptible
to biological attack and even treated rope loses its strength
rapidly underwater. The new synthetic fibers, however, are
immune to biological fouling and decomposition, absorb
little or no moisture, and so are finding increasing use.
Synthetic lines also have the important property of being
highly elastic which makes them ideal for applications where
shock loading and taut-line constraints are present. Creep
can be a problem with synthetics, but with a reasonable
choice of the factor of safety, this is not a serious problem.
Synthetics have low melting points and thus require careful
handling. Certain special braided synthetic lines have been

developed to combat the problem common to all twisted rope and cable, that of untwisting under load. These new lines show great promise and should find increasing applications in the years to come. A problem of unanticipated severity involving synthetic lines is fish bite damage.

One of the problems common to all cables and rope is the limiting length, wherein the weight of line plus attacked equipment becomes equal to the working load of the line. In certain cases, where it is absolutely necessary, tapered cables are used to increase the limiting length. In other cases it is possible to use a synthetic line, such as polypropylene, which is positively buoyant (specific gravity of 0.90) to solve this problem. Properly spaced buoyant devices, such as glass spheres, is another solution. Another common problem is that of strumming; severe vibrations due to vortex shedding when lines are used for towline applications or in strong currents. Premature failure due to strumming can occur at all but the lowest speeds. This problem can be alleviated by the use of snap-on fairings or so-called "haired fairings", but both of these systems are at present extremely costly and further complicate the handling problem.

Connectors and terminal fittings are a continuing source of concern and failures. In large measure, this difficulty is aggravated by the lack of an adequate non-destructive test procedure.

LINE STRUCTURES

Cable and Mooring Systems

The only demand made of the basic ship anchorage is that it be adequate to prevent a vessel from foundering on rocks or beach or colliding with other vessels or shore installations. In most situations this permits very large excursions and even some dragging of the anchor. In general, sufficient chain for 200 feet of water with the normal scope of 5 is felt to be satisfactory. The requirements for floating drilling platforms, drilling rig tenders, buoys for data collection and ocean research stations are much more stringent. To severely limit excursion on station or in depth; to be statically and dynamically stable and to be rugged and long lived are some of the essentials for these mooring systems.

The basic elements of which mooring systems are composed, as illustrated in the figure, are the anchor, line, float and swivels. A taut line used in a simple, single-point mooring reduces horizontal and vertical motions, especially if the float is deeply submerged below surface disturbances. Submergence will also reduce the liklihood of vandalism and loss, but will make location and recovery more difficult at the same time.

A slack line mooring is easier to install and is better suited to absorb shock and surge loadings.

Single-Point Systems

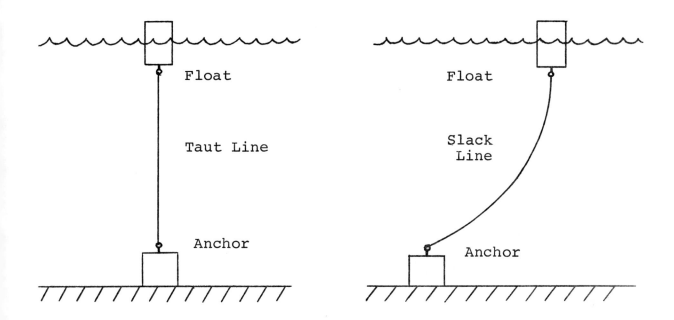

Two-Point Systems

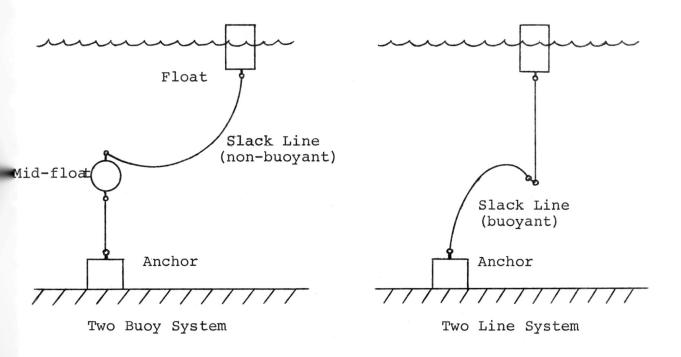

Multi-point moorings are built up in various combinations of these component single-point systems to suit the purpose at hand and they may achieve many degrees of complexity. Lines may be selected so as to be naturally buoyant or not, or they may be made so by external means such as spaced floats.

Compliance with the strictest of motion limitations will require mooring systems of several branches or legs. For ship types, this frequently results in moorings to port and starboard at both bow and stern and one, or perhaps a diverging pair each, forward and aft. For less elongated floating bodies, a plan form arrangement of tic-tac-toe symmetry is natural. Whenever float attachment points are too close together, twisted lines and corkscrewing can result. However, in one instance this problem was avoided by means of a rotatable turret beneath the ship on which converged the eight radially arranged mooring lines.

Counterweights, as well as winches, may be used for tensioning.

Ad hoc, on the spot, self-deployed moorings appear limited to about 1,500 feet of depth at present. Mooring legs tend toward some form of two point or double catenary configuration. Examples of these will be found in offshore drilling and Navy submarine rescue practice.

In the past ten years (beginning with 1958), the U. S. Navy has designed and deployed several deep sea moors for basically scientific purposes. To meet the rigorous and extreme design conditions of the systems, unusual preparations and concentrations of installation ships were required. Several of these "permanent" ship moors have been installed in depths as great as 6,000 feet.

An interesting example is TOTO II. It is probably the first deep sea moor capable of providing both high restraint and firm positioning for large ships. It consists of a center buoy and three others spaced 120° apart with a 1,000-radius. All are tied together by lines 120 feet below the surface. The anchor legs radiate outward, one each from these sub-surface attachment points, in the form of double catenaries to anchors positioned on a 3 1/2 mile diameter circle in 5,500 feet of water. The anchors are set to oppose each other and are connected at the center under 10,000 lbs. tension, producing a firm, flexible structure. The moor was designed for a useful life of five years.

LINE STRUCTURES

Equilibrium Cable Trajectories

In the most general sense, the purpose of a mooring system is to maintain some sort of object in a particular geographical location, either in an absolute sense or relative to some other object. Beyond this basic requirement, however, the exact nature of the object being moored (and its operational requirements) is most likely to dictate the more specific characteristics required, or at least, desired in the mooring system. Once these characteristics are defined, it is possible to consider the problem of optimizing the design of the mooring to produce the system which best meets these requirements. The selection of the cable trajectories is a primary consideration in this process.

The cable trajectory aspect of the design problem is prominent because of the fact that the cable trajectories exercise such a profound influence on the behavior of the cable system. Deriving from the capability of lines and cables to transmit tension but not compression nor flexure forces, such lines at points of attachment must be aligned with the direction along which the greatest force or resistance is desired. Thus it should be possible by judicious selection of the cable trajectories to produce a cable system which has most nearly the desired behavorial characteristics.

While the behavior of a cable system is controlled largely by the cable trajectories, the cable trajectories in turn are controlled by the characteristics of the cable itself, its physical size and shape, its density, and its elasticity, to name the most important. The mechanism of this control is the cable's interaction, either direct or indirect, with the ocean environment, the hydrostatic pressure, the wave motions, the ocean currents, the weather patterns, etc. The buoyancy of a submerged cable and the added mass and drag produced by its relative motion within the ocean medium represent the most important aspects of these interactions. The magnitudes of these effects and the load applied by the cable can be controlled by the proper cable configuration. This in turn is dependent upon the size, cross sectional shape and mass density of the cable. Lightweight buoyancy devices, such as glass spheres, can be spaced at predetermined intervals for final tailoring of the trajectory.

The optimum cable trajectories can be determined only after the design requirements have been defined. These requirements can represent any number of things. A few of the more likely ones are worth mentioning. For an application where the loading is largely static it might be desired to have a cable system which for a given strength level has the maximum strength-to-weight ratio for ease of handling. For cases where the dynamic loads or loads variable in direction are important, minimizing motion, either in terms of distance, velocity, or acceleration might be the prime consideration. This would include those situations where it is desired that shock or excursion be minimized, and in such cases, the elasticity of the cable system would be another property of the cable and trajectory which would be highly significant. Other possibilities might represent weighted combinations of the above.

This discussion originated with the mooring problem. However, it should be obvious that the feasibility of optimizing equilibrium trajectories is not restricted to this problem area. The towing problem, and many others in the area of ocean engineering cable systems, call for application of this technique which, in general, leads to solutions for the tension in the line or its contour length and scope. For simple conditions of no stretch in the line and a uniform current, for example, solutions to the differential equations derived from equilibrium considerations are readily obtainable. However, there is as yet no rigorous or general solution for the more complex problems involving three dimensional moors or compound lines of varying characteristics, varying current profile and dynamic loading. Numerical integration and lumped models must be resorted to.

LINE STRUCTURES

Analysis of Stayed Structures

In a general sense, a stay is a wire or rope line used in some manner to steady or support. In the ocean engineering field, stays are most frequently encountered as supporting elements for masts and antennas. They are used on such structures as a means of preventing excessive bending and/or deflection without resulting in a structure of undesirably large weight or dimensions. Stays may also be essential if the mast step or foundation conditions are such that little or no base end support moment can be applied to the mast.

A simple two-dimensional, symmetrical stayed mast system is illustrated in the first figure. The initial tension in the stays may be denoted by T. If a force system is applied to the mast such that a deflection of attachment point P results, the stays resist such a deflection by means of their elongation. If the small horizontal deflection is δ and the stay

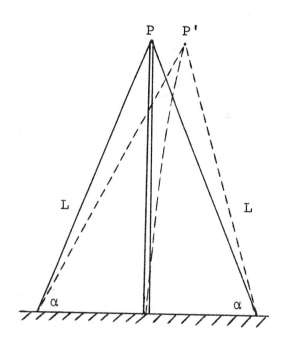

elongation is e, as in the
next figure, it can be seen
that

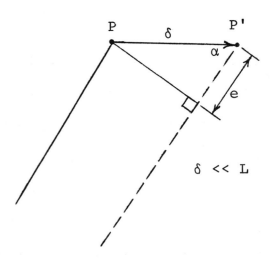

$$\Delta L = e \simeq \delta \cos \alpha$$

Thus **one** stay will increase
in length by the amount ΔL
and the other will decrease
in length by an equal amount.
Such changes in length produce changes in tension in the stays
which can be written

$$\Delta T = AE \left(\frac{\Delta L}{L}\right) = \frac{AE}{L} \delta \cos \alpha$$

where A is the cross sectional area and E is the elastic
constant (Young's modulus) of each stay. The force vectors
at point P' due to the tension
in the stays is shown in the
figures alongside, together
with their horizontal and
vertical resultants. It can
be seen that the result of a
small horizontal deflection
δ is to produce a horizontal
resisting force in the stays,
F_s, such that

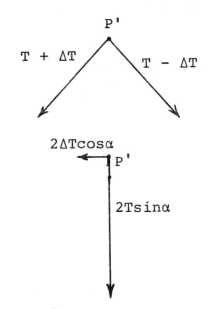

$$F_s = 2\Delta T \cos \alpha = 2 \frac{AE}{L} \delta \cos^2 \alpha$$

Thus, at point P', the top of the mast may be subjected to

three horizontal force components

as shown. Beside F_s there is

F_b, from the shear force in the

mast due to its bending rigidity

and, perhaps, F_o, the applied

loading causing the displacement

PP', and whose existence and

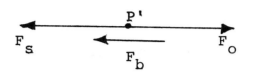

magnitude depend upon the nature of the loading system. (Of

course, the displacing force may be applied at some other

point on the mast.) Since equilibrium requires that the sum

of the forces be zero, imposition of this requirement will

result in a determination of the magnitude of the deflection,

δ. With δ known, the bending moment and deflection distribu-

tions for the complete mast can then be determined.

With the forces applied at P, and no other lateral

load on the mast, for example, the simple cantilever relation-

ship between the end deflection, δ, and the resisting force

in the mast, F_b, is simply

$$\delta = \frac{F_b L_m^3}{3E'I} \quad \text{or} \quad F_b = \frac{3\delta E'I}{L_m^3}$$

where L_m is the mast length, E' its modulus of elasticity

and I its cross sectional moment of inertia. So the

equilibrium equation to be solved for δ in this case is

$$2\frac{AE}{L}\delta\cos^2\alpha + \frac{3\delta E'I}{L_m^3} - F_o = 0$$

An interesting interaction aspect of these problems is that as the stiffness of the mast is increased so does the proportion of the load it carries with the result that there may be little reduction of stress in the mast. This would seem to imply that mast scantlings can be so reduced that some form of primary or secondary form of buckling in the mast might be the more likely form of failure to anticipate in design rather than excessive stresses as such.

While somewhat simplified, this example illustrates the usual approach used to analyze stayed masts, both in two and three dimensions. In the three dimensional case there is the orthogonal deflection component and the corresponding equilibrium force equation with the additional degree of freedom. In addition, the presence of a greater number of stays even from a common attachment point may substantially complicate the problem, both geometrically and computationally.

Other complicating factors may be a variable mast cross section from end to end, swinging booms whose topping lifts apply loadings from any direction (most with components parallel to the axis of the mast), acceleration forces from the oscillations of the system due to seaway motions, and non-symmetries. The stretch of rope in the stays also introduces non-linearities in that its modulus increases with load and with age, so that its value at any particular moment is highly uncertain.

Two points are of particular significance when
working with stayed structures. The first is that the
addition of stays may considerably increase the column load
on the mast. As indicated in a previous figure, the two
stays considered in the simple two dimensional system add a
compressive column force of magnitude $2T \sin\alpha$ on the mast.
The addition of this force not only increases the column
axial stress, but increases the possibility of column
instability. The second point is the possibility of having
slack stays. This can result either from having stays whose
initial tension is zero or from having two large a deflection
in the mast, producing a reduction in tension equal to or
exceeding the initial value. The effect of slack stays is
to effectively reduce the restoring force of the total system since the slack stays can support no compression and
are therefore inactive. Unless slack stays are acceptable,
the initial tension should be determined such that any
anticipated deflection of the mast is not sufficient to
cause any of the stays to go slack.

Weight saving improvements in the system may be
effected by resorting to additional attachment points on the
mast and rearranging stays, perhaps as follows

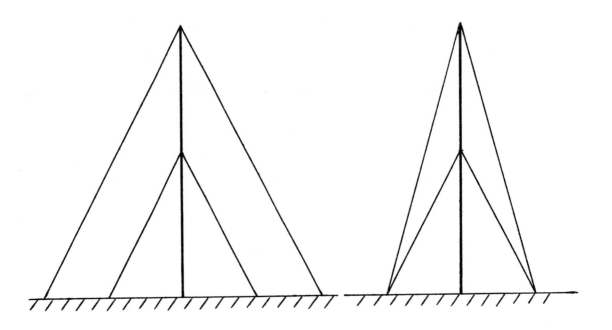

Spreaders are often used in stayed mast design to reduce weight aloft, particularly in cases where the base size for the stays is limited. Such a device is shown in the accompanying figure. Not surprisingly, the addition of spreaders substantially complicates the analysis problem, but a first approxima- tion can be determined by assuming that the part of the overall bending load carried by the mast itself is negligible compared

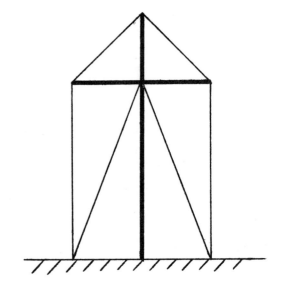

to the part carried by the stays and spreaders. This
approximation is good for very flexible masts, less so
for stiffer masts that support a significant part of the
load themselves. For the "exact" solution, an iterative
solution technique is recommended.

Certain taut line mooring systems in a general
sense can be considered as stayed structures. If the mast
of the first figure is removed and a buoy at P is substituted,
a taut line moor is the
result. In such a system
the initial tension in each
stay is determined by the
buoyancy of the total system, the buoy and the
stays. The trajectories
the stays assume in-
fluence the direction
of the force applied to
the buoy. Consequently, the factors which influence the
trajectories of the stays significantly influence the exact
behavior of the system. The most common primary objective
in the design of such a system is the minimization of buoy
motions under the influence of currents and, perhaps, of
wave forces.

SHELL ANALYSIS AND DESIGN

Unstiffened Cylindrical Shells under
External Hydrostatic Pressure

For marine structures which are designed for
operation at great depths below the surface of the ocean,
the hydrostatic pressure becomes the most critical loading
system on these structures. For this reason it is important
to understand, and to be able to predict the behavior of
structural elements subjected to hydrostatic pressure.

Cylindrical shells, both stiffened and unstiffened,
are very frequently encountered in marine applications, particularly for subsurface work. In the following, the behavior
of unstiffened cylindrical shells loaded by hydrostatic
pressure is discussed with respect to both theoretical and
experimental investigations.

Elastic instability is one of the primary modes of
failure which must be considered. To simplify the mathematics
of the problem, it is usual to treat it as that of the buckling
of a uniformly compressed circular ring. The results are
then modified for the case of the cylindrical shell. The
results obtained with this technique provide very reasonable
accuracy provided one of two conditions is met. The first is
that the shell have free edges. If the edges are not free
then the second condition is that the length, L, of the shell
must be great enough so that the stiffening effect of any

constraint at the edges can be neglected. This is the case
when the ratio, L/D, where D is the shell diameter, is greater
than approximately 20. Assuming that one of these conditions
holds, linear elastic theory is adequate to predict the critical buckling pressures. This theory shows that there are many
critical pressures, each one corresponding to a deformed shape
consisting of a specific number of full sine waves around the
shell's periphery. If k represents this number of full sine
waves the general equation for the critical pressures can be
written

$$P_{cr} = \frac{2}{3} (k^2 - 1) \frac{E}{1 - v^2} (\frac{h}{D_o})^3$$

where E is Young's modulus, v is Poisson's ratio, h is the
shell thickness, and D_o is the diameter to the shell mid-
thickness. The only buckling mode of practical importance here
is the lowest, that corresponding to the value k = 2. This
gives the lowest buckling pressure as

$$P_{cr} = 2 \frac{E}{1 - v^2} (\frac{h}{D_o})^3$$

Experimental evidence indicates that this relationship does a
reasonably good job of predicting elastic instability. However,
this is true only for cylinders with near perfect circularity.
Any slight out-of-roundness has been shown to be capable of
causing degradations of strength on the order of 20 to 30 percent. Thus if the above relationship is to be relied upon to
predict cylindrical shell instability, it is essential that
very close tolerances in circularity be maintained.

A second primary mode of failure which must be considered is that of excessive stress relative to the strength of the material. External hydrostatic pressure will produce in a cylindrical shell a compressive tangential or "hoop" stress and, depending on the nature of the ends of the cylinder, possibly also an axial compressive stress. The magnitudes of these two are the following:

$$\text{Tangential, } \sigma_t = p \times \frac{D}{2h}$$

$$\text{Axial, } \sigma_a = p \times \frac{D}{4h}$$

where the previous definitions apply. The limiting value of stress may be set at a particular value for any of a number of reasons. That combination of axial and tangential stresses which causes initial yielding is one of the most common limits. If the Mises-Hencky yield criterion is assumed, yielding will begin when

$$\sigma_a^2 - \sigma_a \sigma_t + \sigma_t^2 = \sigma_y^2$$

where σ_y is the material's yield stress. Written in terms of the pressure this becomes

$$p_{cr} = \frac{4}{\sqrt{3}} \, \sigma_y \frac{h}{D}$$

From the nature of the instability and yield equations, it can be seen that for a given material each mode of failure will be the controlling mode over a certain range of values of h/D. Thin cylinders, with low values of h/D, will fail by elastic instability; thicker ones, with larger values of h/D, by shell yielding.

Shells in the intermediate range, for which failure occurs at stresses above the proportional limit, will fail by plastic instability at a reduced value of the modulus of elasticity.

The relationship of these failure modes to one another is shown diagrammatically for the case of spheres in the section "Spherical Shells under External Hydrostatic Pressure."

SHELL ANALYSIS AND DESIGN

Unstiffened Cylindrical Shells under Bending and Axial Loadings

In order to consider the problem of cylindrical shells under combined loads, it is at first necessary to understand the behavior of such shells under the action of each load individually. The types of loads in question are external hydrostatic pressure, axial loads, and bending loads.

An axial load, P, on a cylinder of radius, R, and thickness, h, will produce a stress in the axial direction only of magnitude

$$\sigma_{aa} = P/2\pi hR$$

If the cylinder is sufficiently thin, a secondary elastic instability of the walls will occur before this stress gets excessively large. Use of linear elastic theory predicts a critical stress of

$$(\sigma_{aa})_{cr} = 0.60E\frac{h}{R}$$

where E is the Young's modulus of the material and a value of 0.3 has been assumed for ν, Poisson's ratio. Experimental data has shown that this expression gives values of buckling stress which are much too high, a discrepancy which theory has yet to explain. Consequently, recourse has been made to experimental data and an empirical relationship has been determined. This expression is the following

$$(\sigma_{aa})_{cr} = k_c \frac{\pi^2 E}{12(1 - \nu^2)} (\frac{h}{L})^2$$

- 86 -

where L is the cylinder's unsupported length and k_c is an empirically determined coefficient which is a function of the ratio R/h and a dimensionless length, Z_L. Z_L is defined as follows

$$Z_L = \frac{L^2}{Rh} \sqrt{1 - \nu^2}$$

Curves of k_c versus Z_L for discrete values of R/h have been drawn from the experimental data and are available for use with the above relationship.

Once the critical stress has been determined, the critical axial load follows simply:

$$P_{cr} = 2(\sigma_{aa})_{cr} \pi Rh$$

Of course, the Euler primary buckling mode may also be a possibility in some ranges of column proportions.

A moment, M, about a diametral axis will produce an axial stress in a cylinder which varies linearly across the cross section between maximum values of opposite sign at the two extreme points from the axis. The value of the maximum stress, σ_{ab}, is given by the expression:

$$\sigma_{ab} = \pm M/\pi hR^2$$

As with an axial load, if the cylinder is too thin, wall instability can occur at a value of stress lower than the maximum allowable for the strength of the material. In order to make use of data available on buckling from pure axial loads, a "gradient factor", γ is employed which is defined as follows:

$$(\sigma_{ab})_{cr} = \gamma (\sigma_{aa})_{cr}$$

Linear theory gives a value of 1.3 to this factor, and test results seem to substantiate this for cylinders in the so-called "long" range $(Z_L > \simeq 20)$. For cylinders below this range no data are available to permit the recommendation of a value for the gradient factor. Thus at the individual's discretion, the above relation may be used to determine the critical bending stress with a value of γ in the range 1 - 1.3. Once the critical stress is known, the critical moment follows simply:

$$M_{cr} = (\sigma_{ab})_{cr} \quad \pi h R^2$$

For very short lengths, stresses exceeding the strength of the material may be encountered before buckling is likely so the yield mode then becomes critical.

SHELL ANALYSIS AND DESIGN

Unstiffened Cylindrical Shells under Compound Loadings

When considering the case of combined loads, the stresses generated by each of the loads may be simply added. The total axial stress, σ_a, would thus be

$$\sigma_a = \frac{P}{2\pi hR} \pm \frac{M}{\pi hR^2}$$

where the load P includes any axial load generated by the hydrostatic pressure. Hydrostatic pressure, p, also gives rise to a tangential stress component, σ_t, which is given by

$$\sigma_t = \frac{pR}{h}$$

If the magnitude of p is much less than either σ_a or σ_t, its role as a direct stress may be ignored. The Mises-Hencky criterion for initial yielding may be written

$$\sigma_a^2 - \sigma_a\sigma_t + \sigma_t^2 = \sigma_y^2$$

where σ_y is the yield stress of the material, and this provides a satisfactory test of adequacy against yielding.

In treating instability under combined loads, recourse usually is made to the so-called "interaction formula." This relation is of the form

$$\left(\frac{p}{p_{cr}}\right)^{\alpha_1} + \left(\frac{P}{P_{cr}}\right)^{\alpha_2} + \left(\frac{M}{M_{cr}}\right)^{\alpha_3} = 1$$

where the α's are experimentally determined constants. There are two main reasons for the use of such a concept. First, the lack of success in using theory to predict instability behavior for individual loads is not likely to inspire confidence in the ability of the same theory to accurately predict behavior for multiple loads. And second, the form of the interaction formula is ideally suited to curve fitting of experimental data points.

While an insufficient amount of data is available, that which is available seems to suggest that a choice of values for the alphas of

$$\sigma_1 = \sigma_2 = \sigma_3 = 1$$

while being somewhat conservative, is not overly so. This choice results in an elastic instability relationship for cylinders, both stiffened and unstiffened, of

$$\frac{p}{p_{cr}} + \frac{P}{P_{cr}} + \frac{M}{M_{cr}} = 1$$

For the present, in design work it is always desirable to include a factor of safety (F.S.) to cover the many uncertainties of loading. This factor of safety can be included in the shell yielding and elastic instability equations as follows:

$$[\sigma_a^2 - \sigma_a \sigma_t + \sigma_t^2]^{1/2} = \sigma_y/(F.S.)$$

and

$$\frac{p}{p_{cr}} + \frac{P}{P_{cr}} + \frac{M}{M_{cr}} = \frac{1}{(F.S.)}$$

to suit the two possible modes of failure; yielding or instability. It is important in evaluating the axial critical load, P_{cr} , that both primary and secondary instability failures be considered and the most likely (the smallest) value used.

SHELL ANALYSIS AND DESIGN

Spherical Shells under External Hydrostatic Pressure

For pressure vessels operating at great depths, the sphere is by far the most frequently used shape because of its optimal weight /buoyancy or strength/displacement ratio.

As with cylindrical shells, elastic and plastic instability and excessive stress are the most important modes of failure. The elastic instability mode will be considered first.

A linear theory of spherical shell elastic instability was first developed by Zoelly in 1915. This theory gave the critical pressure, p_{cr} as the following

$$p_{cr} = 1.21E\left(\frac{h}{R}\right)^2$$

where E is Young's modulus, h is the shell thickness, R is the radius to the shell mid-thickness and a Poisson's ratio value of 0.3 is assumed. Experimental investigations indicate that this equation greatly over estimates the strength of spherical shells. Virtually no models have been able to survive pressures greater than about 70% of the theoretical value and some have collapsed at values as low as 25%. These results have prompted much additional theoretical investigation in the search for a theory which would more accurately predict collapse, without much success. However, careful examination of the experimental data has produced what appears to be the reason

for the discrepancy between theory and experiment.
Analogous to the case of cylindrical shells, there is much
evidence to suggest that the premature buckling of the
spherical shells occurs because of imperfect sphericity.
As the figures point out, however, this effect is much more
pronounced in spheres than that of non-circularity in cylinders. Any local or gross imperfection such that the local
radius of curvature of the sphere is altered is sufficient to
cause a severe reduction in strength. Consequently, to provide
the necessary level of strength, it is necessary to maintain
extremely close tolerances. In this respect, an empirical
buckling formula has been developed at the David W. Taylor
Model Basin which is a modification of the theoretical. This
equation for the critical pressure,

$$P_{cr} = 0.84E \left(\frac{h}{R}\right)^2$$

can be expected to provide a reasonable estimate of strength
provided that initial departures from sphericity are less than
2 1/2 percent of the shell thickness.

An external hydrostatic pressure will generate a
uniform state of compressive stress in a thin sphere. The
magnitude of this compressive stress, σ, is given by the
relation

$$\sigma = \frac{1}{2} p \frac{R}{h}$$

Setting the limiting stress at the value of the yield stress allows for the solution of the critical pressure for failure by shell yielding. This value of pressure is

$$p_{cr} = 2 \sigma_y \frac{h}{R}$$

where σ_y is the value of the yield stress.

For a given sphere of a given material, the failure mode which gives the lowest value of the critical pressure will be the controlling mode. For small collapse depths, and hence low values of the ratio h/R, instability controls. Above a certain value of h/R the shell yielding mode of failure assumes control. Consequently, for a sphere of a given material a graph of collapse pressure versus h/R will have the form of the curve below.

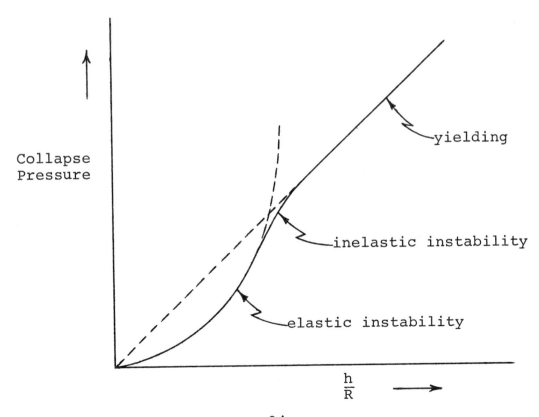

The figure which follows indicates the relative
performance of spheres of various materials, based upon their
shell weight/buoyancy ratios. For purposes of comparison,
this ratio of W/B was limited to 0.33 to allow for weight
allocations for power source, controls, life support,
instrumentation, etc. The symbol ρ_s is the mass density of
the structural material and ρ_w is that of sea water. Despite
differences in specific weight, strength and elasticity from
one material to another, it is evident from their collapse
depths how the materials rank.

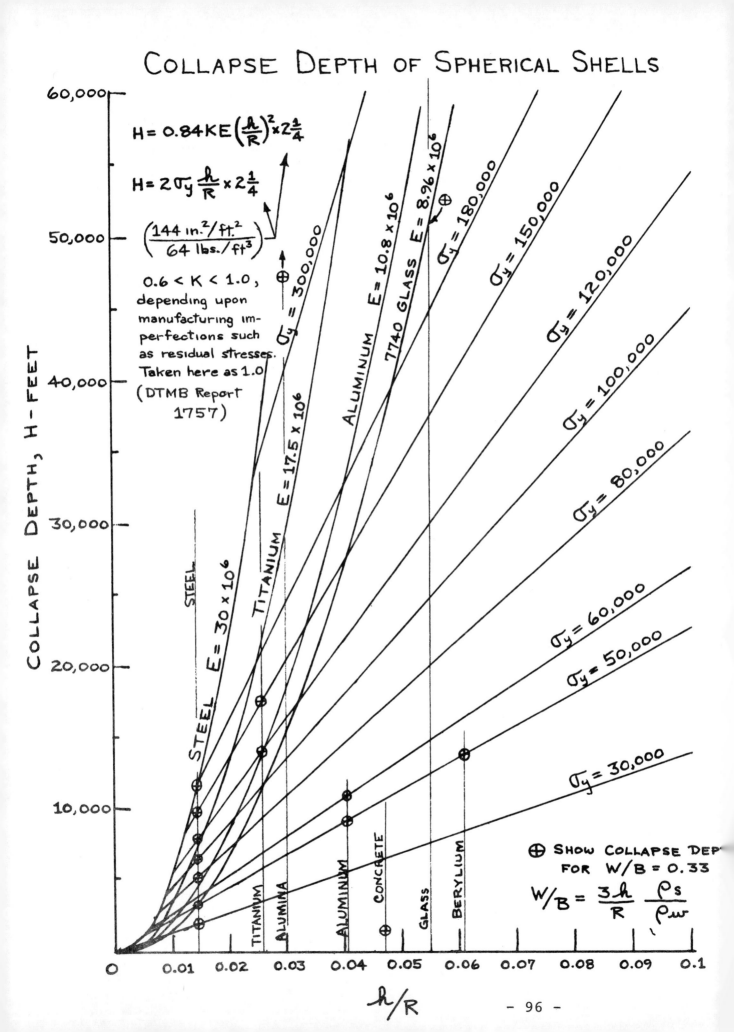

SHELL ANALYSIS AND DESIGN

Direct Design of Shells

It may be postulated that in a shell structure
under load, its optimum shape (or more correctly its shape
for minimum weight) will be achieved when at all points
on the surface and through the thickness the resultant
stresses are equal and at the maximum allowable stress level.
This implies a membrane stress distribution, or in other
words, no bending across the thickness. By "direct design"
is meant the process by which such a shape and thickness
variation may be determined for a given load definition, in
a manner both rational and systematic. Alternatively, the
objective may be the minimum enclosed volume or the least
cost. Or perhaps it is required to predict the loaded shape
of a variable thickness membrane structure, such as an
inflated bladder.

On the other hand, there may be certain ruling
geometrical constraints, such as a partially constricting
hard tank, which prevents completely attaining the uniform
stress condition, but for which the shape of the unconstrained
portion is desired. Instead, it may be the resultant stress
distribution in the uniform thickness, but partially con-

stricted envelope which is desired.

Some work has been done on generalized solutions
to such problems for axisymetric cases by means of numerical
methods. Because of the iterative nature of these solutions,
it has been convenient to resort to automatic machine
computation, for which modern computers are very suitable.
Shallow shells have been investigated because the reduction
of the equations to Laplacian form, with the introduction of
stress functions, permits an inversion between stress distributions and geometric parameters.

In one case a pure membrane state of uniform stress
resultants is specified. The independent variables are the
geometric parameters. The solution is then the geometry of
the shell in which the stresses satisfy the membrane
equilibrium equations. Within such a limitation, the pro-
grammed procedure can be made to converge on the least weight
solution. The initial point must be specified by two geometric
boundary conditions resulting in an initial value problem.

In the other case, constraints on the geometry may
be specified by requiring the shell to pass through a
number of fixed points (or in reality, circumferential rings).
The stresses will be no longer uniform or constant but will
be made to vary in such a manner that the geometric conditions
can be met. With one geometric condition specified at
each of two boundaries the problem becomes a boundary value

problem.

It is also possible to specify only one of the principal stresses independently. The other will be a function of the first and of the geometrical properties of the shell.

Typical of the problems for which the method is uniquely suitable are rigid and non-rigid pressure vessels inherent in operations in the oceans and in which the form restriction to surfaces of revolution is of no consequence. To illustrate its validity; when required to define the shape of the buoyancy chamber of a submerged buoy tethered on taut lines to the ocean bottom, a series of onion shapes was produced explicitly proportioned and contoured to suit the relative magnitudes of external pressure versus the vertical equilibrium force.

SHELL ANALYSIS AND DESIGN

Design of Stiffened Cylindrical Shells
under Hydrostatic Loading

Until recently, the only shell structures
employed in the ocean deeps were the submersibles.
Consequently, it is here that the engineering experience
is the most extensive. Heretofor, submarines have been
put to military use almost exclusively so that the design
literature is largely classified and therefore generally
unavailable. There is however, a second source in the
American Society of Mechanical Engineers, and similar
foreign agencies, whose committees have been engaged in
formulating codes for boilers and unfired pressure vessels.
Historical Review - The pressure hulls of early submarines
conformed to the external molded lines of the vessel and
thus varied in cross sectional shape from end to end.

With the ever-increasing efficiency of detecting
devices and anti-submarine weapons, greater operating and
collapse depths have been desired. Furthermore, the search
for a more efficient structural whole has continued with the
result that the modern naval submarine is radically different
in several respects. The pressure hull throughout the
midship portion is generally cylindrical and conical with

spherical or elliptical end closures and, as such, has the
sections of inherently greatest strength as, ideally,
only axial or membrane type loadings are involved and
not bending.

Being entirely self sufficient for long periods
of time, with large payload demands and high speed
requirements, it follows that compared to the new develop-
mental generation of research submarines, naval types are
much larger and tend more toward elongated rather than
spherical pressure hull shapes. Initially, depth capabilities were modest. Shallow depth designs being volume
limited, and with demands being made for housing ever
greater quantities of equipment and gear, many items were
moved outside this strength hull; fuel oil and ballast tanks
in particular. Thus, generally the pressure hull became
an inner shell having the minimum of undesirable, abrupt
shape changes due to "hard" tanks, with a lightweight
outer shell maintaining the fair form amidships and merging
with the single hull which still prevailed at the ends.
Minimizing the pressure hull size is clearly advantageous
in saving overall weight.

The optimum hull form for underwater propulsion
permits the use of circular sections even to the extreme
ends of the ship so that now, when long endurance nuclear

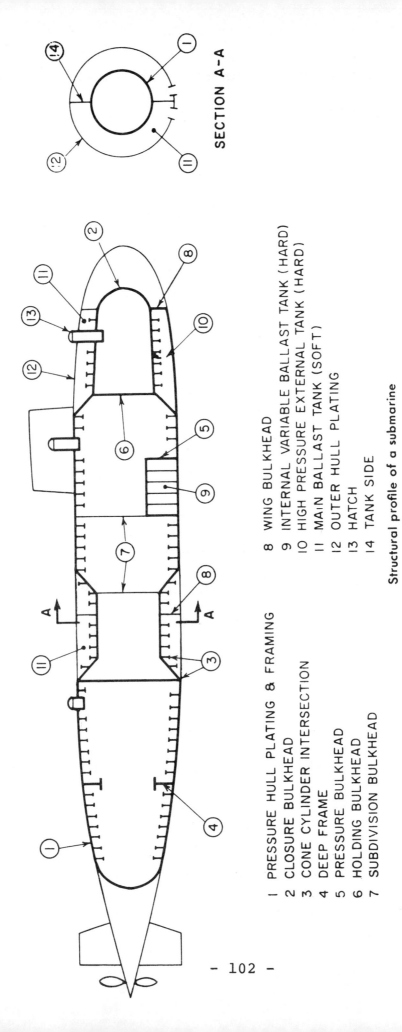

SECTION A-A

1 PRESSURE HULL PLATING & FRAMING
2 CLOSURE BULKHEAD
3 CONE CYLINDER INTERSECTION
4 DEEP FRAME
5 PRESSURE BULKHEAD
6 HOLDING BULKHEAD
7 SUBDIVISION BULKHEAD
8 WING BULKHEAD
9 INTERNAL VARIABLE BALLAST TANK (HARD)
10 HIGH PRESSURE EXTERNAL TANK (HARD)
11 MAIN BALLAST TANK (SOFT)
12 OUTER HULL PLATING
13 HATCH
14 TANK SIDE

Structural profile of a submarine

From "Principles of Naval Architecture" by J. P. Comstock, S.N.A.M.E. Chapter IV, "Strength of Ships" by D. F. MacNaught.

power plants are used and there is little need for surface operation, noncircular shapes have almost entirely disappeared.

With a nuclear power source, smaller tankage is required so that "single hull" construction now obtains over a greater percentage of the ship's length than ever before.

The circular frames of most present day submarines are welded and in way of "double hull" construction they may be located outside the inner or pressure hull, providing additional unobstructed space inboard. Space saving and weight saving are also apt to result in external frames with sea floor habitations as well.

Failure of Cylindrical Shells under External Pressure

Stiffened and unstiffened cylindrical shells are considered to have failed when, under external load, a circumferential accordian pleat develops or their circularity in a transverse plane is destroyed and a major series of lobes or bulges has formed circumferentially to such an extent that the structure can be said to have collapsed.* This ultimate

*As a corollary definition, it may be postulated that "failure" has occurred when an abrupt decrease in pressure hull volume has taken place, exceeding the vessel's capacity to restore the neutral buoyancy condition before it has submerged too deeply to recover. An appreciation for this possibility may be necessary for proper interpretation of experimental model tests performed in tank systems in which, unlike the sea, stored energy is not limitless and the structure may unload itself in deforming.

failure may result from any one of several localized
origins or be of several types. The most likely of these
are (1) elastic and plastic instability of the shell whereby it buckles between the frames while the frames remain
circular, (2) yielding of the shell, and (3) general instability of the shell and frame in unison. These will now
be considered in this order in somewhat more detail. Ideally,
for optimizing strength on a weight basis, it is a widely
held belief that the structure should be so designed that
failure by as many of the three modes as possible is expected
to occur simultaneously, i.e., at the same pressure.

Instability of the Shell - The advent of shell buckling is
evidence that a portion of the structure has shirked its
load, and transferred it elsewhere, and complete collapse
may be imminant as a result of such triggering action.
Once the shell has buckled, it can contribute little more
to the strength of the structure. The pressure at which
buckling occurs is dependent upon the length, diameter
and thickness of the tube in addition to the physical properties of the material with the assumption being made that
the length of the cylinder is the distance between supports
which simply maintain the circularity of the section, but

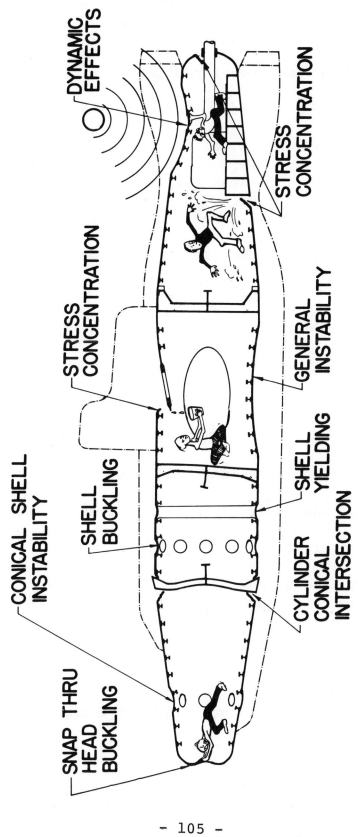

Significant Modes of Structural Failure

(Courtesy of the Naval Ship Research and Development Center)

do not necessarily furnish any other constraint. When the

supporting rings are infinitely far apart, they exert no

influence in strengthening the shell, and failure takes

the form of elementary two-lobed instability collapse

at relatively low pressures. In decreasing the ring

spacing beyond this "critical length", it will be found that,

in addition to the thickness-diameter ratio, the length-

diameter ratio now takes effect in strengthening the vessel.

The closer the spacing the greater the load the shell will

sustain before buckling and the larger the number of lobes

formed (which mode is still one of elastic and finally,

plastic instability). Ultimately, a second transition

point is reached beyond which decreasing the spacing of

supports has less effect and, to all intents and pur-

poses, only the thickness-diameter ratio is again uniquely

important. In this third region, stresses approaching the

yield point occur in the shell.

The theoretical formula for elementary "two-

lobed" collapse* of a long thin tube under radial pressure

has been given by Breese and Bryan as--

---

*Actually, four lobes or two waves are formed.

$$P_{cr} = \frac{2E}{1 - \nu^2} \left(\frac{h}{D_m}\right)^3 = 66 \times 10^6 \; (h/D_m)^3 \text{ for steel.}$$

Well-known expressions for the elastic instability failure of thin tubes within the region where stiffer spacing or length has an effect upon the collapse pressure have been formulated by Southwell, von Mises, and Tokugawa. In their complete form, they are cumbersome and involve n, the number of waves formed circumferentially upon buckling, which is also an unknown.

From the most correct and accurate of these, Windenberg of the U.S. Experimental Model Basin developed the following expression which is independent of the number of lobes and gives results averaging only about one percent different from its ancestor within the range of the majority of practical cases; viz.,

$$P_{cr} = \frac{2.42E}{(1 - \nu^2)^{3/4}} \quad \frac{(h/D_m)^{5/2}}{\left[L/D_m - .45 \; (h/D_m)^{1/2}\right]} \text{ or,}$$

for steel,

$$= 2.60E \; \frac{(h/D_m)^{5/2}}{\left[L/D_m - .45 \; (h/D_m)^{1/2}\right]} \quad \text{on the assumption that}$$

$$n = \frac{\pi D_m}{L}$$

---

*For definitions see Nomenclature

From the above, and the figure, it is apparent
that the spacing of stiffeners is now a factor of
importance.

With small L/D_m ratios, it becomes impossible to
get an accurate picture of the failure of the stiffened
cylinder without considering the contribution of the ring
stiffeners in themselves carrying part of the load. Nevertheless, in attempting to proceed rationally and most
efficaciously to a total design solution, and having as
yet no notion of the stiffener characteristics, it is useful temporarily to overlook their contribution of area and
reckon the hoop stress on the basis of the plating alone.
Although maintaining circular the frame line of the shell,
it might be expected that when the compressive hoop stress
exceeds the yield point, the load which the shell will
carry is no longer its full share of the total and, because
of plastic deformation, removal of the load is not accom-
panied by a return of the material to its initial state and
configuration. Such a condition is unacceptable. This
stress limitation, when incorporated in the familiar hoop
stress formula, gives the following equation:

$$P_{cr} = 2(h/D)\sigma_y$$

$$P_c = 2\left(\frac{h}{D}\right)\sigma_y$$

$\sigma_y = 80,000\ \text{psi}$

$\sigma_y = 50,000\ \text{psi}$

A

1

3

Equal
Weight
Solutions

2

$$P_{cr} = \frac{2.60\,E\,(h/D_m)^{5/2}}{\frac{L}{D_m} - 0.45\left(\frac{h}{D_m}\right)^{1/2}}$$

$h/D = 0.0042$

$h/D = 0.0031$

D = 20 ft.
h = 1.00 in.
h = 0.75 in.

$\sigma_y = 50000\ \text{psi}$
&
$\sigma_y = 80000\ \text{psi}$

$$P_{cr} = \frac{2\,E}{1-\nu^2}\left(\frac{h}{D_m}\right)^3$$

Unsupported length/Diameter (L/D_m)

Collapse pressure P_c – psi

Collapse depth – feet

which, though dealing in terms of the yield stress appears to closely approximate the most likely mode of failure which might better be described as plastic instability.

The figure has been drawn for two constant values of h/D in such a manner that the plot of each formula is shown only in the area where it applies and with the transitions empirically defined. A plot of such curves for a number of h/D values is useful in making a first approximation to the required pressure hull thickness and the frame spacing. An examination of the figure given will show that at least from the consideration of these curves alone, the most advantageous position for the choice of pressure hull scantlings is on the upper "shoulder" contour of one of the curves slightly to the right of such a point as the tangent point "A". Obviously, a closer spacing of frames than at "A" does not increase the collapse pressure of the vessel despite the additional number of stiffeners and probable weight increase involved. Based on this diagram then, points anywhere on the horizontal line to the left of the shoulder do not represent the best balance of shell and stiffener scantlings. By slightly increasing the frame spacing over that at "A" and assigning the frame weight saved toward increasing the shell thickness, a stronger vessel would result. This will be the case until the point under consideration is located too far over on to the sharply sloping elastic portion of the curves as illustrated by the points for equal weight solutions 1, 2 and 3. Practical usage would probably place design points conservatively to the left of the "shoulder" in order to comply with the tolerances to which plates are rolled and further considerations. Also, it is probably well to recall that in the foregoing expressions, the stiffeners have not been given any credit whatever for any portion of the total load they may carry. This approach then, while helpful in understanding, leads only to a first, tentative conclusion for shell thickness and frame spacing, given a required collapse pressure and shell diameter.

Frame Spacing - Stiffened cylindrical shell test data were collected by Windenburg and Trilling and plotted on a nondimensional basis in terms of ψ and λ.

Such a plot is given in the following figure and it is helpful in selecting a frame spacing, as the large number of experimental data points in the low lambda region makes it possible to locate a "design line" in that area quite definitively.

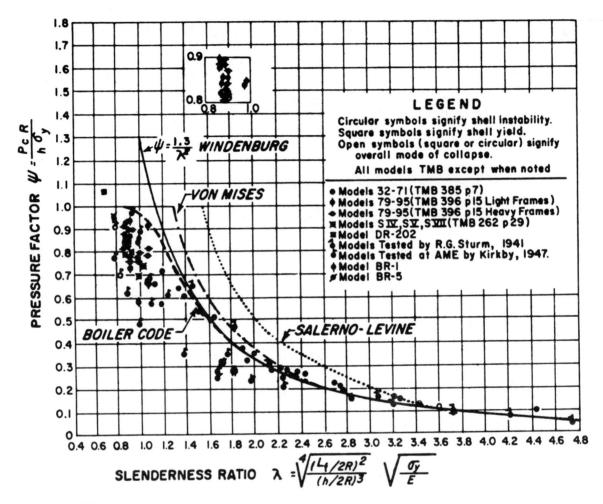

From "Pressure Analysis of Submarine Hulls" by
Edward Wenk, Jr., <u>Welding Journal</u>, June 1961,
p. 275-s.

On this figure only

    L_f = Unsupported span of plating, in.

    R   = Radius to neutral axis of shell alone, in.

In view of the previous discussion, ψ values
of about unity are desired and the corresponding value of
λ is thus taken to be in the neighborhood of 0.80.

The relatively smaller frame spacing and larger
numbers of frames associated with increasing strength of
steel, increases the proportionate load taken directly by
the frames. Correspondingly, it may be expected that the
pressure necessary to collapse the shell is greater than that
which induces a simple compressive circumferential hoop
stress (calculated by excluding frame area) of yield strength
magnitude.

The external frame location offers inherently
greater strength. The explanation is that internal frames
are especially sensitive to the effects of unwanted initial
tilt in bringing about tripping of frames under load, whereas
with external frames the hydrostatic load on the shell oper-
ates to reduce any initial tilt there may be and restore the
frame to its intended normal orientation.

Frame Strength - Not only on account of direct weight economy
are overly heavy frames undesirable. In addition, they may
overplay their part and because of excessive rigidlty, actually
cause premature failure of the shell by inducing in it additional components of stress. The limit to be sought then

is how weak may the frame rings be and still be adequate.

In early submarine design, a variation of the Foppl or Levy formula for the instability collapse of a circular ring under uniform external pressure was used; viz.,

$$p_c = \frac{24EI_b}{D_b^{\,3}L_1}$$

In practice, a factor $\frac{1}{1.1}$ was incorporated to insure that the stiffener be an arbitrary ten percent stronger than the shell, giving:

$$p = \frac{p_c}{1.1} \quad \text{or}$$

$$p = \frac{1}{1.1} \times \frac{24EI_b}{D_b^{\,3}L_1}$$

The hope was to provide ample margin against failure so that the frames could support the entire external pressure load even after the shell had buckled and then contributed nothing except as the narrow band of material in way of the frame was effective in augmenting the moment of inertia of the section.

Rewritten, this expression becomes:

$$I_b = 0.046 \frac{D_b^3 L_1}{E} p$$

and with the approximation, $\frac{I_b}{I} = 1.3$, established from an analysis of usual stiffener and plating scantlings then in use, tentative frame scantlings might be chosen in a straight-forward manner.

Leading to a lighter frame is the Tokugawa frame formula expressed here in the form for direct comparison with other frame formulas:

$$P_c = \frac{24EI_1}{D_m^3 L_1} \left\{ \frac{1}{1-\nu^2} \left[\frac{1}{12} L_1 h^3 + L_1 h \left(\gamma - \frac{h}{2} \right)^2 \right] + \right.$$

$$\left. \left[I + A \left(h - \gamma + \mathfrak{z} \right)^2 \right] \right\}$$

$$= \frac{24EI_1}{D_m^3 L_1} \times \frac{1}{1 - \nu^2}$$

Here the width of shell plating considered in the compound moment of inertia is equal to the frame spacing.

Von Sanden and Gunther developed an expression for the loading, usually known as "Formula #88", which when incorporated into the Foppl formula contributes another frame formula which, when $\nu = 0.3$, has been expressed as:

$$P_c = \frac{24EI_b}{D_b^3 b \dfrac{1 + \dfrac{0.85\beta}{B}}{1 + \beta}}$$

where $B = \dfrac{bh}{A + bh}$

$$\beta = \sqrt{\frac{11N}{100\,\dfrac{h}{D_m}}} \quad x \quad \frac{h^2}{A + bh}$$

and $N = \dfrac{\cosh\theta - \cos\theta}{\sinh\theta - \sin\theta}$

where $\theta = 10 \quad \sqrt[4]{12(1-\nu^2)} \quad \dfrac{\dfrac{L}{D_m}}{\sqrt{100\,\dfrac{h}{D_m}}}$

$$= \frac{18.2\,\dfrac{L}{D_m}}{\sqrt{100\dfrac{h}{D_m}}}$$

The following figure presents the salient features of an early report on the results of experiments made to examine the validity of these frame formulas. Although the experimental models were confined to a small range of characteristics, the figure itself is instructive in calling attention to the relationship between various formulations and modes of failure. Also the experimental points do indicate a wide margin of overdesign in the "standard practice" formula then in use and apparent conservation in the formulas of Tokugawa and von Sanden and Gunther as well.

From "The Influence of Stiffening Rings on the Strength of Thin
Cylindrical Shells under External Pressure" Charles Trilling,
U.S.E.M.B. Report #396

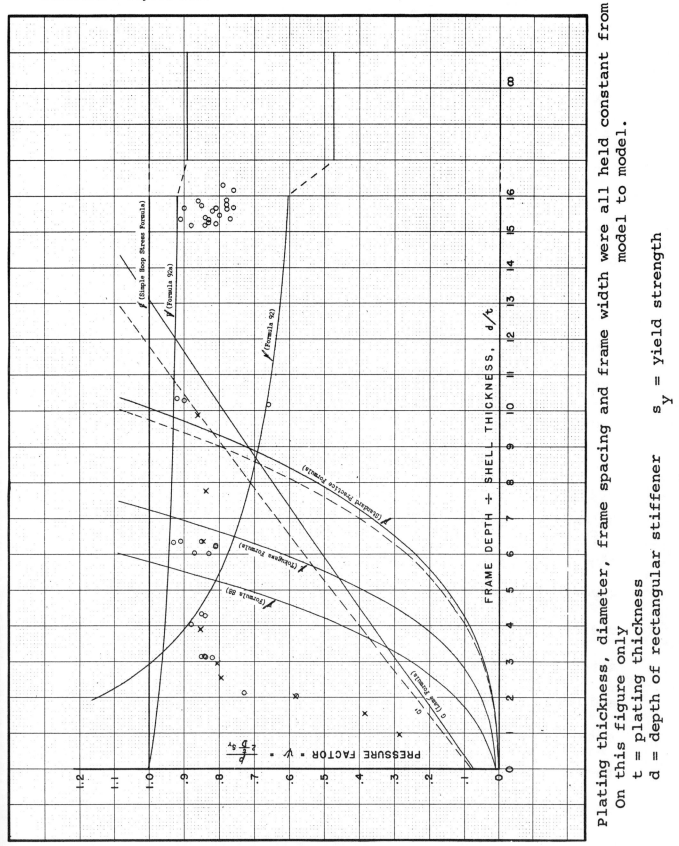

Plating thickness, diameter, frame spacing and frame width were all held constant from
model to model.
On this figure only

t = plating thickness

d = depth of rectangular stiffener s_y = yield strength

GRAPHICAL REPRESENTATION OF THEORETICAL FORMULAS
AND EXPERIMENTAL DATA.

It is now generally recognized that a stronger, more resilient type of construction is that in which frames and shell are equal in strength as opposed to the hard-framed structure typical of early practice. The excess rigidity of the frame bringing about stress concentrations in the shell might, therefore, better be reduced to aid in obtaining the high structural efficiency required of submarine pressure hulls. If frame weight could be cut in the process, it could be used to better advantage as additional shell thickness, thereby increasing the collapse pressure of the structure still further.

Effect of Frames on Shell Plating Strength - The first approximation to the shell plating thickness for a close spacing of frames may be made from the hoop stress formula as already discussed, but subsequent selection of frame scantlings makes possible a more thorough stress analysis on the basis of additional formulas of Von Sanden and Gunther. Commonly known as "#92" and #92a", these expressions are for the external pressure at which stresses in the shell reach the yield strength, at which point failure was anticipated; #92 considering the stress fore and aft, (i.e. parallel to the cylinder axis) at the edge of the stiffener faying flange

on the inside of the shell, and #92a being concerned with
the circumferential or tangential stress at a point midway
between the frames on the outside of the shell. It may be
well to observe that, fundamentally, in a simple cylindrical
shell, the circumferential stress is twice the axial stress.

Formula #92 (axial stress):

$$P_c = \frac{2\frac{h}{D_m} \sigma_y}{\frac{1}{2} + 1.81 \ K\frac{0.85-B}{1+\beta}}$$

Formula #92a (circumferential stress):

$$P_c = \frac{2\frac{h}{D_m} \sigma_y}{1 + H \frac{0.85-B}{1+\beta}}$$

where B, β, and θ are as defined earlier and

$$K = \frac{\sinh \theta - \sin \theta}{\sinh \theta + \sin \theta}$$

$$H = -\frac{3 \sinh \frac{\theta}{2} \cos \frac{\theta}{2} + \cosh \frac{\theta}{2} \sin \frac{\theta}{2}}{\sinh \theta + \sin \theta} \quad \text{approx.}$$

The previous figure includes solutions to these
two equations and while very often they give concurrent values

of p_c that actually are more nearly equal, the plots do show that as the strength of frames is increased, failure of the shell may occur at progressively lower pressures. To a some-what lesser degree experiments and experience with realistic cross sections and more normal frame spacings bear this out.

The experimental points make another point; viz., given at least adequate frame strength, the cause of ultimate collapse in the plating of a thin-walled shell is excessive circumferential stress rather than longitudinal stress and there may be excessive yielding of the shell at the toes of frame flanges before collapse finally occurs due to high circumferential stress or perhaps to some entirely different failure mode. However, as a caution, if the necessity for making repeated dives to the neighborhood of the maximum operating depth is anticipated, the matter of fatigue may be of concern in submarine design.

Yield Failure of the Shell - Von Sanden and Gunther were in fact applying the Rankine or "maximum stress" theory when they supposed that failure would follow, once the largest principal stress reached the yield strength of the material (assuming failure by another mode had not occurred previously,) As has been discussed, the outcome of this reasoning is not entirely supported by tests because principal stresses in

- 120 -

the shell first reach the yield strength at the stiffeners.
Yet in general, considerably greater pressures than those
causing this stress can be carried without failure. True,
the Von Sanden and Gunther circumferential solution, as one
of the two possibilities, does yield accurate predictions in
one range of geometric proportions but a doubt remains that
with extremely thick plating greater reliability might be
achieved by means of a more acceptable failure theory, such
as the Mises-Hencky "maximum strain energy" theory. The
question here is not with the prediction of stresses, but
with the manner in which stresses are employed to predict
failure.

The experimental findings certainly suggest that
failure at the panel boundaries alone (i.e., analagous to the
formation of hinges) does not bring about a failure mechanism. Mid-bay failure creating a third hinge is also
required. The hinges will be produced in the order mentioned,
ultimately making the mid-bay location critical.

For the two-dimensional case, the Mises-Hencky
criterion may be written:

$$\sigma_1^2 - \sigma_1 \sigma_2 + \sigma_2^2 \leq \sigma_y^2$$

where σ_1 and σ_2 are the two principal stresses in the plane of the plating. In terms of the submarine cylindrical pressure hull and the proposed hypothesis of Wenk, Stark and Peugh calling attention to the mid-bay, mid-thickness stress condition, this is:

$$\sigma_T^2 - \sigma_T \sigma_F + \sigma_F^2 \leq \sigma_y^2$$

The Von Sanden and Gunther Formulas #92 and #92a rest on the assumption that pressures and stresses are linearly related at all stresses up to the yield strength. Thus it is legitimate, for example, to write Formula #92a in more general terms, viz.

$$p = \frac{2 \frac{h}{D_m} \sigma}{1 + H \frac{0.85 - B}{1 + \beta}} \qquad \text{or, alternatively}$$

$$\sigma = \frac{D_m p}{2h} \left(1 + H \frac{0.85 - B}{1 + \beta}\right)$$

mid-way between frames, in the outer surface.

Both #92 and #92a also result from the superposition of bending stresses upon membrane stresses which latter are given as

$$\sigma_F = \frac{D_m p}{4h}$$

at all points, and

$$\sigma_T = \frac{D_m p}{2h} \left(1 + 2Q \frac{0.85 - B}{1 + \beta} \right)$$

mid-way between frames, where

$$Q = - \frac{\sinh\frac{\theta}{2} \cos\frac{\theta}{2} + \cosh\frac{\theta}{2} \sin\frac{\theta}{2}}{\sinh\theta + \sin\theta}$$

Incorporating these values of σ_F and σ_T for a particular set of structural characteristics into the failure condition leads to the following expression for the collapse pressure, based on failure by shell yielding.

$$P_c = \frac{2h\sigma_y}{D_m} \times \frac{1}{\sqrt{F^2 - \frac{F}{2} + \frac{1}{4}}}$$

where

$$F = 1 + 2Q \frac{0.85 - B}{1 + \beta}$$

If this discussion applied to shells with very widely spaced stiffeners, in which case $\sigma_T = 2\sigma_F$ exactly, an examination of the failure equation reveals that failure according to this hypothesis would not occur until $\sigma_T = 1.16\sigma_y$. In other words collapse would be delayed until the pressure was 16% greater than that given by the simple hoop stress formula, or a ψ value well in excess of one.

Lunchick has developed a method which also relies on the Mises-Hencky theory of failure as applied to the shell mid-way between the frames. It too conceives of a three hinge mechanism of failure and explicitly treats the formation of the critical mid-bay hinge as an elastic-plastic development with increasing pressure. His method is difficult to use except by means of the following figure. It yields a factor which relates the collapse pressure (per Lunchick) to the pressure at which yielding first occurs in the outer surface at mid-bay according to the Mises-Hencky criterion (with biaxial stress estimates according to Von Sanden and Gunther). Lunchick calls this latter the "yield pressure".

To find the factor, ratios of circumferential bending to membrane stress and longitudinal to circumferential

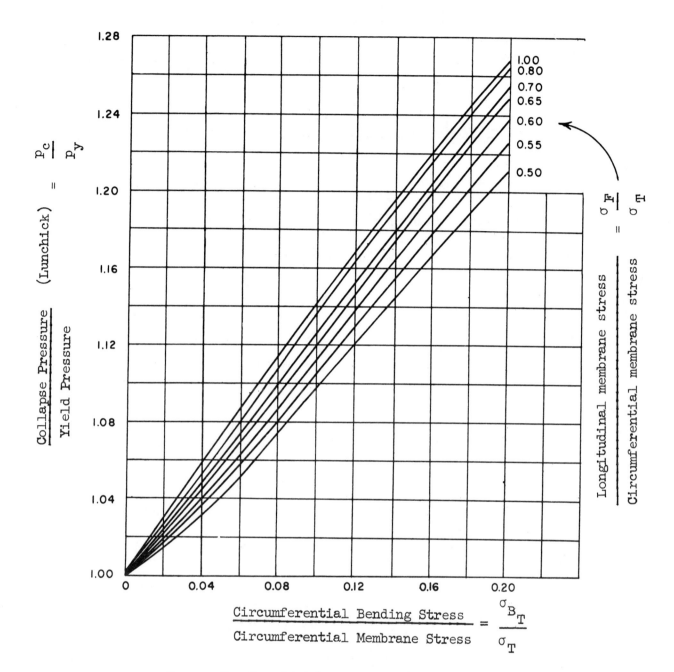

Curves for Lunchick's Collapse Pressure Factor

From D.W. Taylor Model Basin, Report 1291

membrane stress (all at mid-bay) are required. The Von Sanden and Gunther expressions for each of these in turn are as follows:

The circumferential bending stress,

$$\sigma_{B_T} = \frac{D_m p}{2 h} T \frac{0.85 - B}{1 + \beta}$$

where

$$T = - \frac{\sinh \frac{\theta}{2} \cos \frac{\theta}{2} - \cosh \frac{\theta}{2} \sin \frac{\theta}{2}}{\sinh \theta + \sin \theta} \quad \text{approx.}$$

The bending stress at mid-bay is:

$$\sigma_{B_F} = \frac{D_m p}{h} 1.81 \frac{0.85 - B}{1 + \beta}$$

so the total longitudinal stress is:

$$\sigma = \frac{D_m p}{h} \left[\frac{1}{4} + 1.81 \ T \ \frac{0.85 - B}{1 + \beta} \right]$$

When the total circumferential and total longitudinal stresses are involved in the Mises-Hencky failure relationship, the following expression for the "yield pressure" results:

$$p_y = \frac{2h \, \sigma_y}{D} \frac{1}{\sqrt{x^2 - xz + z^2}}$$

if

$$X = (1 + H \frac{0.85 - B}{1 + \beta})$$

and

$$Z = (\tfrac{1}{2} + 3.62 \, T\frac{0.85 - B}{1 + \beta}) = 2(\tfrac{1}{4} + 1.81 \, T \frac{0.85 - B}{1 + \beta})$$

The Lunchick results were also in good agreement with test results.

General Instability - General instability is the term given the instability failure which involves shell and framing as a single entity collapsing with the formation of large lobes between nodal points enforced at points of greater transverse strength, as in way of bulkheads or heavy web frames. Superimposed buckling of the shell between frames may or may not take place.

A solution to this problem was found by Kendrick whose first work dealt with "general instability" of the form illustrated at the top of the next figure.

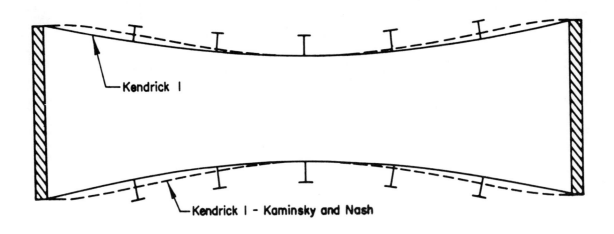

Kendrick I

Kendrick I - Kaminsky and Nash

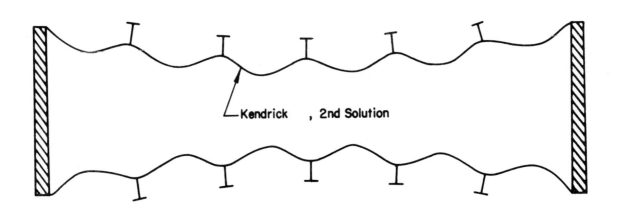

Kendrick , 2nd Solution

Comparison of Buckling Shapes Assumed by Kendrick, Nash, and Kaminsky

Bryant has converted the results of this first solution of Kendrick's into a simplified readily usable form which was believed to differ from the original by no more than 5 percent in the usual range of interest. Bryant's form is--

$$P_{cr} = \frac{2Eh}{D_m} \frac{m^4}{(n^2 + \frac{m^2}{2} - 1)(n^2 + m^2)^2} + \frac{8(n^2 - 1)}{D_m^3 L_1} EI_1$$

I_1 may be approximated by--

$$I_1 = \frac{A(\xi + \frac{h}{2})^2}{1 + \frac{A}{L_1 h}} + I + \frac{L_1 h^3}{12}$$

"n" in this case is the number of circumferential waves formed by the unstiffened plating alone for the length between bulkheads and is not the same number as "n" in the previous expressions which are for plating alone buckling between frames. Strictly speaking, both "n" and "m" can only be integers. An appropriate value for "m" is given by $\frac{\pi D}{2\ell}$ where ℓ is the bulkhead spacing, or the distance between other such points of exceptional transverse stiffness. For the usual submarine structural characteristics, "n" is 2, 3, or 4. Theoretically, the first critical mode of failure (i.e., the one for which the critical load is smallest) is the value sought. The values of "n" and "m" used actually should be those which produce this first critical load. The values given above are estimated to be those leading to this result.

The two terms of Bryant's expression have separable physical significance which is an advantageous feature for design usage when it is desirable to be able to assess readily the effects of changes and so to optimize the total structure. In effect, the formula states that the critical pressure for a ring-stiffened cylinder in a mode involving "n" circumferential lobes is equal to the sum of the buckling pressure in the same mode for the shell alone (supported only by its end bulkheads) plus the buckling pressure for a ring composed of a typical frame with one frame space of its associated plating. Bryant likens this to two Euler columns acting in parallel.

Probably, the most significant oversimplication in this expression is the neglect of cross product terms arising from interaction between the two effects accounted for.

The second phase of Kendrick's work considered inter-frame plate buckling such as studied by von Mises and also illustrated in the lower part of the figure. The implication is, of course, that upon failure several lobes may develop circumferentially in the plating between frames in association with another pattern of lobes forming involving plating and frames together.

Kendrick's second solution covers the possibility of these two configurations being coincident and yields collapse pressures that are less than those given by his earlier work.

Effects of Imperfect Circularity on General Instability -
The stiffened shells discussed in the previous sections were assumed to be perfectly circular, a condition unfortunately never attained. It follows then that with any out-of-roundness some variation of bending moment circumferentially is introduced where none would exist otherwise. Associated with this, of course will be a stress variation through the depth of the stiffener at any section, whereas a uniform stress would otherwise have been the case. Under external loading, there will then be a circumferential strain variation causing radial deflections that tend to amplify the original shape imperfections. Increasing the load increases

the stresses and the pattern of deformation nonlinearly,
until finally collapse occurs. This progressive growth of
initial imperfections has been substantiated experimentally
and emphasizes the extreme desirability of obtaining perfect
circularity as nearly as possible, especially with the frames.

All this, of course, is analagous to what takes
place in perfect versus imperfect or eccentrically loaded
columns. Attempts to define the failure load in the latter

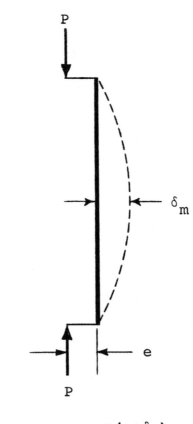

$$\sigma_{max} = \frac{P}{A} \pm \frac{P(e+\delta_m)c}{I}$$

usually do so in terms of the maximum stress occurring in the

extreme fibers of the beam and are composed of two terms;

one, the P/A stress, giving the nominal stress level and the

other, a maximum, extreme fiber bending stress which is

superimposed.

But δ_m is a function of P, increasing from zero as

P increases. δ_m thus introduces a nonlinearity, and σ_{max}

is not simply variable with the first power of P.

It is possible to account for this nonlinearity,

and in about the relationship conforming to the facts, by

introducing Southwell's approximate eccentricity correction

factor, $\dfrac{P}{P_{cr} - P}$, or $\dfrac{1}{\dfrac{P_{cr}}{P} - 1}$ in which P is the actual load

on the column, and P_{cr} is the critical load for the column

ideally straight. Then the equation can be written--

$$\sigma_{max} = \frac{P}{A} + \frac{Pec}{I} \; \frac{1}{\dfrac{P_{cr}}{P} - 1}$$

Obviously, when the load is small, the only eccen-

tricity provided for is "e" whereas when the load approaches

the critical, the second term blows up and yields a stress

indicative of failure. In fact, any resultant stress greater
than the yield strength of the material is presumed to result
in failure.

Relating this line of reasoning to circular rings
of uniform cross section and expressing results in terms of
the external pressure gives--

$$\sigma_{max} = \sigma_C + \sigma_B$$

$$= \frac{2RLp_1}{2Lh} + \frac{2RLp_1\delta}{2} \times \frac{h}{2} \times \frac{12}{Lh^3} \left(\frac{p_1}{p_{cr} - p_1} \right)$$

$$= \frac{P_1R}{h} + \frac{6p_1R\delta}{h^2} \left(\frac{p_1}{p_{cr} - p_1} \right)$$

For one section of a stiffened shell including one
frame and its associated plating, σ_C for a frame space unit
becomes--

$$\sigma_C = \frac{p_1R}{A_1} L_1$$

But the best definition of the loading on the plate-
frame combination is believed to be that given by Von Sanden
and Gunther and referred to previously as Equation #88. The

- 134 -

width of plating involved in cooperating with the frame and in transmitting pressure to the frame directly was taken as b. The total load on the frame (including shear in the shell at the toes of the frame) is found to be--

$$p_1 b R \left[\frac{1 + \frac{0.85 \beta}{B}}{1 + \beta} \right]$$

and then

$$\sigma_C = \frac{p_1 R}{A_b} b \left[\frac{1 + \frac{0.85 \beta}{B}}{1 + \beta} \right]$$

Based on the theory of minimum potential energy, Kendrick obtained a solution for the bending term which can be approximated as--

$$\sigma_B = \frac{Ec}{R_1^2} (n^2 - 1) e \left(\frac{P_1}{P_{cr} - P_1} \right)$$

The complete expression for σ_{max} is then,

$$\sigma_{max} = \frac{P_1 R}{A_b} b \left[\frac{1 + \frac{0.85 \beta}{B}}{1 + \beta} \right] + \frac{Ec}{R_1^2} (n^2 - 1) e \left(\frac{P_1}{P_{cr} - P_1} \right)$$

- 135 -

This stress occurs in the frame flange as the neutral axis is certain to be nearer the shell. Supposing a maximum out-of-roundness can be estimated for the fabricated vessel, and with tentative scantlings of plating and stiffener at hand, the maximum fiber stress can be found for matching against an allowable stress, such as the yield strength, in the usual structural test for satisfaction.

NOMENCLATURE

A   Cross sectional area of frame, in^2.

A_b Cross sectional area of frame plus plating of width b, in^2

A_1 Cross sectional area of frame plus plating of frame spacing width, in^2

b   Width of frame flange in contact with shell, in.

c   Distance from neutral axis of bending to the extreme fiber, in.

D   Diameter to outside of shell, in.

D_b Diameter to neutral axis of frame plus plating of width b, in.

D_m Diameter to neutral axis of shell alone (i.e., mid-thickness), in.

E   Modulus of elasticity, lbs/in^2

e   Eccentricity, in.

h   Thickness of shell, in.

I   Moment of inertia, frame alone, in^4

I_b Moment of inertia, frame plus plating of width b, in^4

I_1 Moment of inertia, frame plus plating of frame spacing width, in^4

L   Unsupported span of plating, in.

L_1 Unsupported span of plating taken as equal to the frame spacing, in.

ℓ Unsupported span of stiffened shell between points of large transverse strength (such as bulkheads), in.

n   Number of complete waves in the buckled configuration, taken in the circumferential direction.

P   Axial load, lbs.

p   Design collapse pressure, lbs/in^2

p_c Predicted collapse pressure (whatever the mechanism) lbs./in.^{2}

p_{cr} Critical buckling pressure, lbs/in.^{2}

p_1 Actual pressure, lbs./in.^{2}

R   Radius to outside of shell, in.

R_1 Radius to neutral axis of frame plus plating of frame space width, in.

γ Distance from neutral axis of compound section to the plating, in.

δ Deflection, in.

δ_m Maximum deflection, in.

ν Poisson's ratio

ξ Distance from faying surface to neutral axis of frame cross section, in.

σ Normal stress, lbs./in.^{2}

σ_B Bending stress, lbs./in^2 (Linearly variable)

σ_C Axial stress, lbs./in^2 (Uniformly distributed)

σ_T Tangential (or circumferential) mid-thickness stress, lbs./in^2

σ_F Axial mid-thickness stress, lbs./in^2

$\sigma_{max.}$ Maximum normal or direct stress, lbs./in^2

σ_y Yield strength, lbs./in^2

SUBSURFACE HABITATIONS, OBSERVATORIES
AND PRODUCTION UNITS

Functional Requirements

With advancing technology, both the feasibility and
desirability of subsurface laboratories and habitations
steadily increases. There follow some of the scientific,
industrial, and military aspects of such laboratories and
dwellings, along with brief references to man's limitations
in the sea.

Oil Production - Present drilling operations off the continental shelves, one of the last unexplored petroleum reserves,
have been especially rewarding but, with the present technology, production has seemed generally unfeasible and unprofit-
able. Many feel this is due to the use of dry land techniques underwater. The idea has been proposed for the use of
subsurface dwellings to service whole fields of well heads in
near shore installations which would be almost completely
free of surface conditions. Conshelf and the Sealab experiments have demonstrated the feasibility of such arrangements.

Fish Farming and Ocean Agriculture - With the problem of an
ever increasing population, increasing reliance on the
ocean as a source of food is inevitable. As on dry land, the
application of advanced scientific techniques should be able
to increase the oceans' yield of food materials. The subsurface laboratory would allow scientists to work on virtually
all of the problems related to this field.

Geological Surveys - From a fixed subsurface laboratory, men
could work on the ocean floor and subfloor in their immediate
vicinity. Detailed geological surveys could be made of small
areas and the use of underwater vehicles would greatly extend
these areas. The use of undersea laboratories would allow
exact relocation of survey sites and the day to day planning
of work and hence would more effectively use the diver's time
in collecting and surveying and less of it in compression and
decompression time.

Prospecting - Unmanned underwater systems can explore large
areas and detect magnetic, thermal and other properties of
materials. They can also collect and analyze bottom material
as well as photograph or televise the ocean floor. At least
for the near future, it does not appear that subsurface
laboratories will be used for prospecting for commercial
purposes, especially as surface methods have already disclosed
rich finds as yet untouched.

Harvesting Mineral Resources - As with oil production, the use
of subsurface dwellings would allow the operations to be free
from the disturbance of surface conditions, would extend the
diver's useful work time, and would allow continual operation and maintenance under all weather conditions.

Scientific Research - There are innumerable problems in many
fields of science for which the availability of manned undersea laboratories and habitations would be desirable and/or
necessary; in connection with the phenomena of sea floor
spreading, ocean circulation, human physiology, archeology,
etc.

The presence of man at depth in the sea is both
a great asset and a limiting factor. Three important problems
are encountered in extending man deepter into the sea: (1)
physical, mechanical problems; (2) toxic effects of chemically
active substances; and (3) the management of medical facilities to aid incapacitated divers at depth. The first two
must be considered as major problems and some specific items
in these categories are mentioned below. Most of these
arise because of the abnormally high pressures and cold
temperatures of the ocean environment. Inert gases under
high pressures are responsible for the "bends" and for narco-
sis which may cause serious impairment of both mental and
physical capabilities. Oxygen under too high a pressure
may be toxic, and the process of breathing under high pressure
requires greater effort. In addition, large hydrostatic
pressures have derogatory effects on a number of biological
functions. The cold temperatures of the deep ocean tend to
predispose an individual to develop the bends and since body
heat is lost approximately 21 times faster than in air, his
endurance is considerably shortened. The difficulties are not
insurmountable, and existing knowledge should provide for
extension of manned laboratories and dwellings at this time
to depths of 600 feet. Tank tests dives to over 1000 feet
have already been made.

Relative weightlessness is another complication and
special means must be provided for performing even the simplest
of manual operations.

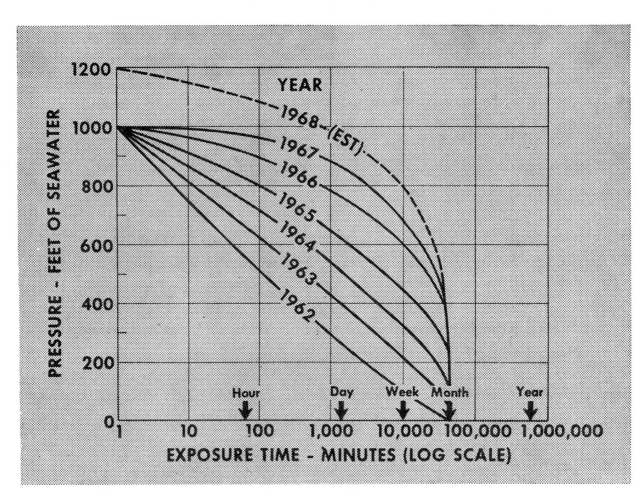

Ocean Industry, August 1968
Volume 3, No. 8, - pg. 28

DIVING TRENDS

SUBSURFACE HABITATIONS, OBSERVATORIES
AND PRODUCTION UNITS

Design Case Histories

A few experimental ocean floor structures have
been built and tested, but at present there are none
emplaced on a permanent or semi-permanent basis with a
specific long range objective in mind. The scientific and
economic potential of the ocean and the ocean floor being
recognized, much thought is currently being devoted to the
possibilities, the result being that many ideas have been
advanced on potential structures for the ocean floor.

One of the ideas being considered is the use of
pressure equalized flexible bags for the underwater storage
of liquids such as oil. Because the bags would have no
bending stiffness they would behave as interface membranes
and the internal pressure of the stored liquid would be the
same as the external water pressure. This would eliminate
the need for a pressure hull which for underwater structures
is almost always the primary structural consideration.

For storage of materials other than liquids and
for manned ocean floor habitations, the use of pressure
equalized flexible bags is less attractive. Some sort of
anchored and bottom supported rigid pressure vessel is the
more common type of ocean floor structure and the manned
experimental habitats which have been used have almost all
been of this type. For moderate depths and greater, such

structures usually employ a pressure hull of stiffened cylindrical or spherical configuration or some combination of these two. The working internal pressure of such pressure hulls, depending on function, may be anything between sea level atmospheric pressure and the pressure of the surrounding ocean. Instead of the gaseous mixtures, as in habitations, ocean bottom oil storage tanks are likely to have the unused storage space occupied by sea water admitted through bottom ports in response to the varying oil storage demand.

Materials suggested for such structures include the more usual metals, glass, cement, and the fiber reinforced composites. The idea has also been proposed that such ocean floor structures be constructed in modular units which could be interconnected to form a complete unit of a desired configuration and size. This would enable the structure to be tailored to a specific application as far as internal volume and arrangement is concerned, while at the same time employing standardized units.

Several variations of the previous category of pressure vessels for applications involving work on the ocean floor have been suggested. These might be termed bottom penetrating pressure vessels. Such vessels would have an open bottom which would be set down on and into the ocean floor. In this way, the ocean floor itself would form one barrier to the external environment. Once such a vessel was in place, water would be evacuated from within the vessel to be replaced with a life support atmosphere. Thus work could be carried out on the ocean floor from within the structure without the need of diving gear. Probably the most critical problem with such an arrangement is that of the seal preventing the external sea water from leaking or seeping back into the vessel. The severity of this problem would depend on the characteristics of the soil of the ocean bottom, the depth of submergence, and the pressure of the internal atmosphere. At a given location the first two are fixed, so that control of this problem can only be exercised by regulating the internal pressure. It remains to be seen whether such measures are capable of providing a practical solution while simultaneously allowing for reasonable working conditions.

Crucial to the use of sea floor containment vessels is their installation and emplacement. If they are temporarily sealed off for this purpose, the choice of whether to arrange for small negative or positive surface buoyancy will depend largely on whether lowering is to be done from some surface support, subject to seaway motions or drawn to the bottom by means of fittings or motive devices which may be difficult to emplace. Differential compressibilities between sea water and the vessel and the increasing net force attributable to paid out handling gear must be anticipated. Clearly, this aspect of the problem will be accentuated with a gas filled vessel if pressure equalization is permitted. On the other hand, the soft containment structure of the pressure equalized vessel will undoubtedly weigh less than its counterpart, the hard tank.

Bottom currents and time varying loadings within the vessel can demand stays and, hence, more positive anchoring devices for a greater variety of bottom conditions than otherwise.

A high rise, buoyant, stayed flotation chamber a few hundred feet below the surface could provide a reasonably accessible, relatively low pressure, stable platform from which many tasks might be performed, on the bottom and above, in great depths of water.

Sealab I - Sealab I was a four man dwelling which was submerged in 195 feet of water off Argus Island for twenty-one days in July of 1964 at the initiation of the U. S. Navy. It was cylindrical in shape, 40 feet long and 10 feet in diameter. Sealab I used a helium-oxygen gas mixture at bottom pressure. Access was through an open access trunk in the bottom of the cylinder. Railroad axles of 300 pounds each were used to obtain the necessary negative buoyancy and lowering was done using a 9 inch nylon line, a system which resulted in large oscillations while lowering. In addition, it was found difficult to maintain sufficient internal pressure during lowering and so to prevent flooding. This problem has plagued other, even smaller and less complex devices.

Sealab II - Sealab II was larger than its predecessor, being 57 1/2 feet long and 12 feet in diameter. It was designed for 10 men and was submerged for 30 days at a depth of 205 feet in August 1965 off LaJolla, California. Some important changes were made in the design of Sealab II in view of the difficulties encountered with Sealab I. Sealab II was designed to be fully pressurized before being lowered and was designed according to ASME codes for non-fired pressure vessels. A thickness of mild steel of one inch resulted. An obvious but significant fact here is the more critical aspect of the internal rather than the external loading. Water was used as variable ballast instead of the axles used on Sealab I. Access while on the surface was by means of a conning tower; but when submerged this tower was flooded, and access was then provided by means of an access trunk on the bottom of the cylinder as with Sealab I.

A more sophisticated, counterweighted lowering system was used for Sealab II which functioned satisfactorily. When placed on the bottom, it was resting with about a 6 degree angle in both directions. This proved troublesome because it prevented complete emptying of the ballast tanks, and thus made the raising operation more difficult.

Sealab III - The Sealab III experiments are now scheduled for early 1969. The same habitat as for Sealab II will be used, but two rooms below the main structure have been added. One is a diver-changing station and the other an observation

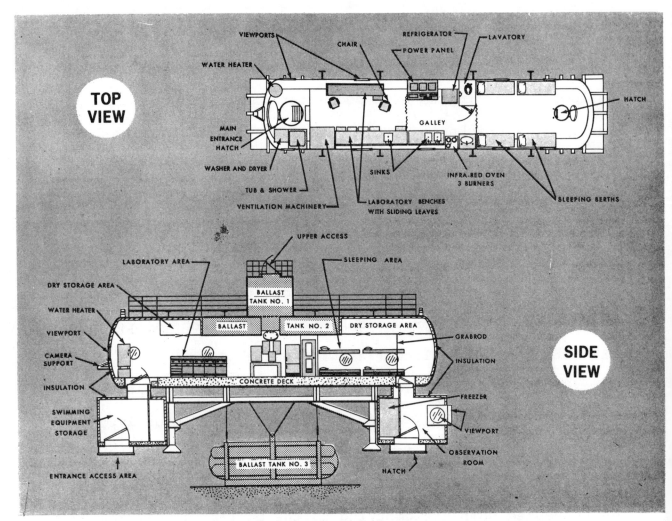

SEALAB III interior schematics.

"SEALAB III - Giant Step Toward Occupation of the Sea Floor"
by Rear Admiral Philip D. Gallery, USN (Ret.), <u>OCEAN INDUSTRY</u>
Oct. 1968, Vol. 3, No. 10.

compartment. The habitat will be placed in 600 feet of water off the California coast and with its surface support vessel is expected to be on location for 60 days. Five teams of eight men each will participate, each team spending 12 days on the bottom. Experiments planned involve testing man's physical as well as mental behavior and experiments in underwater construction, marine biology, geology, acoustics and the evaluation of new instruments, tools and equipment - such as heated suits for the "aquanauts".

Conshelf Experiments - Since 1962, Captain Jacques Cousteau has been carrying on a series of experiments to test man's ability to perform useful work in the undersea environment and over prolonged periods of time. Two weeks' time submerged has been the usual target time, but the average overall elapsed time has been more nearly a month. As the series has progressed, depths have increased to 328 feet for the dwelling unit with Conshelf Three in 1965. The endeavor has also been to reduce dependence on surface life support.

A number of structures has been used for a variety of auxiliary purposes in addition to the habitat. In place on the ocean bottom they are pressure equalized, and the main unit has become spherical in shape.

Flexible Structures - Edwin A. Link has designed and operated two types of flexible dwellings. One of these, the SPID (Submersible, Portable, Inflatible Dwelling) was fabricated from a commercially available rubber storage bag. The bag was mounted in a pipe frame to hold ballast, gas cylinders, and to provide a lifting point. The SPID was successfully operated with 2 men at a depth of 432 feet for a period of 49 hours. The other device, IGLOO, was a bottom work area for surface divers which consisted of a nylon reinforced rubber hemisphere with a diameter of 8 feet attached to an external ballast ring. This device was operated at a depth of 40 feet for 50 days.

SURFACE OBSERVATORIES AND INSTRUMENT PLATFORMS

Design Requirements and Design Case Histories

The category of surface observatories and instrument platforms could include floating structures from simply instrumented surface buoys to research vessels and the new spar type floating platforms. The functions performed by these devices cover a wide spectrum and coupled with such other factors as cost, availability of equipment and materials, exercise considerable influence on their design.

For situations where observations and measurements are to be performed at a fixed location over a substantial period of time, and where the quantity of equipment and personnel necessary are not great, instrumented buoys may be suitable. Buoys have the advantage of being relatively inexpensive and, if not too large, they are not handicapped by a lack of mobility. Because of their being unattended for lengthy intervals, mooring failures have been a problem. Buoys have very frequently been lost. For more elaborate observations and measurements, and where the presence of men and elaborate equipment are required, a ship will be needed. Research vessels provide much more in the way of facilities, such as on the site workshops and laboratories, facilities for man-in-the-sea operations and storage of specimens; in addition to almost unlimited mobility and the ability to respond quickly to changes in prevailing conditions. These advantages come at greater expense however, both in terms of initial and operating costs. In addition, because

of the complex motions of such a platform, obtaining

sufficiently reliable data may be an uncertain or more

costly undertaking on account of the data reduction and/or ship

stabilization required. The limited budget o most oceano-

graphic operations and the considerable laboratory and

equipment facilities necessitated by a diversity of tasks,

add unduly severe space requirements to the lengthy list of

demands which the naval architect must customarily reconcile.

As a kind of compromise between buoys and ships,

there are the spar type platforms. These elongated devices

are usually towed to the site in the horizontal position.

On site they are set in the vertical position by ballasting,

so that about 80% of their length is beneath the water surface.

This large draft, together with a very small waterplane area,

produces a platform whose dynamic response is almost completely

decoupled from wave action at the surface. This results in

platforms which are extremely stable with respect to magni-

tudes of pitch, heave, and roll motions; a very desirable

feature for surface observatories. This unique arrangement

also produces some unique problems. Hydrostatic stability

must be provided for the vessel in all attitudes of operation,

in the horizontal, the vertical, and in transition. In

addition, during this progression from horizontal to verti-

cal positions, stability with respect to rotation about the

longitudinal axis is also likely to be required.

In the transition mode the structural loads can be severe; being comparable to those produced under tow among waves. Large bending loads arise which, together with the large length to depth ratios of spar platforms, make the longitudinal strength problem a major design consideration. Because of the major role they play during the transition period and in tuning platform motions, the ballasting provisions also become a major design consideration. And finally, with the platform assuming such extremely different attitudes, the question as to the choice of integrated versus demountable end-mounted laboratory units arises. Integrated facilities require that most furniture and equipment be mounted on gimbals or somehow easily reoriented and resecured. Demountable units avoid this complication by being placed on and off the platform while it is in the vertical attitude only, but there is the obvious handling problem.

On account of their lack of significant motion in a seaway, spar type platforms may become even more desirable as bases for penetrating the air-sea interface, for example, as in casting instruments adrift or launching and retrieving divers or submarines from below. Hydrostatic loadings will undoubtedly control the structural design of the immersed end of spar platforms. This leads naturally to cylindrical configurations and transverse ring stiffeners. During transit or up-ending, longitudinal bending is critical

so that longitudinal framing is to be expected in the
neighborhood of the mid length where the bending stresses
are maximum. Such a framing system is likely to be more
well developed in the upper portion, as shell thicknesses
demanded by the larger external pressures at the lower
end will be less subject to instability failure.

To provide a stable platform with a low background
noise level for making underwater acoustic measurements at
sea, the Federal Government and industry have built several
"spar" type platforms. To date, all such devices are towed
to the working location. Features of three such craft follow:
FLIP (FLoating Instrument Platform)
General Dimensions
Length = 355' Draft Vertical = 300'   Draft Horiz. = 10'
Diameter = 20' at stern, narrows to 12.5'
Towing speed: Normal = 8 knots, Max. = 11 knots
Displacement = 1500 tons (towing) 2000 tons (vertical)
FLIP, as with all spar type platforms, because of its very
large length to depth ratio had a longitudinal strength
problem. This required a large transverse section modulus
resulting in plating 1/2" thick in the 20 foot diameter sec-
tiontion and 3/4" thick in the 12.5 foot section. In
addition, distributed liquid loading was used to decrease
longitudinal bending stresses.when among waves.

Because of the large draft in the vertical position,
resistance to external hydrostatic pressure was provided by

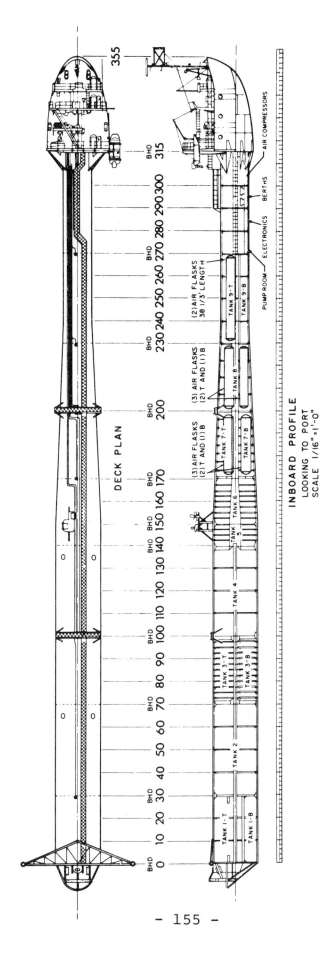

FLIP (Floating Instrument Platform)

"Ocean Engineering" by John F. Brahtz, John Wiley &
Sons, Inc., New York, 1968.

circular cross sections, radially uniform scantlings and transverse stiffening frames. Due to the increase of pressure with depth, frame spacing varied from ten feet to thirty inches.

A small waterplace area, coupled with deep draft in the vertical position, resulted in a platform with very small wave response either in heave or pitch/roll. Sixty tons of concrete ballast were added to improve the stability primarily in the transient flipping operation.

The ballast system's primary purpose was to control flipping. There were free flooding tanks, "hard" tanks, trim tanks, and air tanks; all but the free flooding tanks being built to withstand the external sea pressure. The concrete ballast provided a righting arm for prevention of rolling the vessel over about its longitudinal axis when up-ending or returning to the horizontal position.

At the forward or upper end is space for a laboratory and living accommodations which had to be made suitable for use in both orientations. (In a similar device devised by Mr. Cousteau, this space was designed as a modular unit to be lowered in place after flipping.)

SPAR (Seagoing Platform for Acoustic Research)

General Dimensions

Length = 354' Draft Vertical = 300' Draft Horiz. = 10'

Diameter = 16' throughout

Towing speed = 6 knots

Displacement = 1373 tons (towing) 1738 tons (vertical)

SPAR presented its designers with the same type of problems as did FLIP. Unlike FLIP, however, SPAR has a constant diameter throughout which eliminated the problem of structural discontinuities. Steel with a yield stress of 33,800 psi was used which resulted in a plating thickness of 1/2 inch throughout. As with FLIP, SPAR has its shell stiffened transversely and longitudinally although the latter is more for ease of construction than for strength reasons.

SPAR is unmanned and data is transferred by cable to a tender ship.

POP (Perpendicular Oceanographic Platform)

General Dimensions

Length = 233'

Diameter = 5'9" at bow, 3'6" at stern

Displacement = 60 tons horizontal, 115 tons vertical

POP is somewhat smaller than SPAR or FLIP. In order to reduce the weight of the upper (stern) section, no longitudinal stiffeners were employed which resulted in a rather limber structure. This condition added critical stresses at the transition zone of the diameter. To provide the necessary strength, 3/4 inch piping was used.

Unlike SPAR and FLIP, POP has a platform at the vertical end onto which a trailer, instrument unit is lowered after POP has been placed in the vertical position. This platform can support up to 6000 pounds. In addition, there is a decompression and observation chamber, accessible from the top, used for divers to exit or enter.

OFFSHORE DRILLING AND PRODUCTION PLATFORMS

Design Requirements

Offshore work platforms provide, for people living above the ocean surface, the capability of working on and below the ocean surface and the ocean bottom. The functions such platforms may be called upon to perform are many. They may serve as platforms for prospecting and production of oil and minerals, platforms for programs of scientific research, weather stations, and as manned navigation aids to name the more likely ones. These functions, along with the environmental conditions the platforms will face while both on and off station, provide the designer with a wide variety of design considerations to serve as a basis for the decision making process. In general, it is the most important and common design considerations which are enumerated and discussed briefly in the following.

In some applications it is either highly desirable or absolutely necessary to limit platform excursions, roll, heave, and/or pitch motions to levels within certain tolerable values. Such motion stability limits call for a floating platform whose motions in the above modes are strongly decoupled from wave and wind action. Furthermore, positioning capabilities, especially as applied to horizontal translation and heading, may be necessary so as to achieve and maintain a precise geographical location and orientation. In certain instances it is essential that horizontal motion be kept to

an absolute minimum, for example, so as not to excessively
strain the rotating drill pipe in the case of a floating
rig over a drill site. In such cases, sophisticated and
activated anchor cable mooring systems or dynamic mooring
systems (wherein propelling devices control positioning by
means of position sensors) are required; the depth of water
being a prime factor in choosing between the two. With
shallow water in relatively undisturbed locations, simple,
passive anchor cable mooring systems may be adequate.

When some type of bottom supported platform is to
be used, the characteristics of the ocean bottom at the
location in question will significantly influence the design.
The magnitudes of the loads which can be supported by the
bottom soils, without settlement, must be determined. If
piles are to be used, the depth to which they would have
to be driven to provide sufficient support must be known, and
the ability to drive piles to the necessary depth with the
equipment and techniques available must be assessed.

The failure of platforms in the past must continue
to induce a re-examination of survivability, of the design
loadings, and a reassessment of design criteria. In transit
conditions may be most critical.

The term "removability" refers here to the ease
with which a bottom supported platform can be disassociated
and removed from one operating site so that it can be operated
at another. In some cases the complete platform may be re-

moved from the site. In others, only parts of the platform
may be removed; certain sections of a permanent nature being
abandoned at the site. Silting in and bottom suction in
soft sediments are the principle factors to be overcome.

Scour is, in general, the movement of sediments
by water action. Scouring around the base of a bottom
supported rig has caused differential subsidence and ultimate loss of the structure.

Strong winds are an abvious threat, either for the
wind loadings themselves or because of the waves and currents
they generate. If a platform has to be towed or propelled
in the open ocean it must have sufficient hydrostatic and
hydrodynamic stability, along with adequate structural
strength, to withstand the rigors of the sea without sustain-
ing any of the many forms of failure, such as buckling of
structural members or capsizing and sinking of the complete
platform. A major source of loss has been open ocean transport.

The influence of economics is obvious, but it is
included here because of the large variety of systems from
which one may be chosen and because the cost of the different
types varies so considerably. Even so, as in all complex
systems, it is the total of life cycle costs which are
important; not the acquisition cost alone. Merely as an
illustration of one of the many vagaries which may be
involved; permanent, bottom supported platforms may be

economically justified in deeper water off the coast of
California than elsewhere, where unlike the Gulf of Mexico,
production is not limited by government regulated depletion
allowances. And for almost each global region there is a
regionally important loading which may be governing; hurricanes, scouring and bottom subsidence in the Gulf of Mexico,
earthquakes off the coast of California, ice in Cook Inlet.
etc.

OFFSHORE DRILLING AND PRODUCTION PLATFORMS

Structural Types

Offshore work platforms have been developed in a variety of sizes and shapes and degrees of complexity. However many features are found in common among the different platforms which has led to a number of schemes for classifying them, including the amount of equipment on the structure, the degree of mobility and method of installation, and the method of support, to name three. These considerations can result in a very fine breakdown, but for the purpose of this discussion platforms will be grouped into two major categories: bottom supported platforms and floating platforms. Implied in this categorization is an increasing operating depth capability and hence a chronological order.

BOTTOM SUPPORTED PLATFORMS

Stationary Platforms - Stationary platforms are used for production and commercial development of a site where a platform of a more permanent nature is desired. In deeper water these platforms usually consist of a "template", which is a three dimensional framework or jig which rests on the ocean bottom and through whose legs piles are driven and made integral. Templates are customarily floated to the site and sunk or launched from a barge. Early offshore operations in very shallow water made use of platforms erected on piles only. Stationary platforms have been either self-contained units,

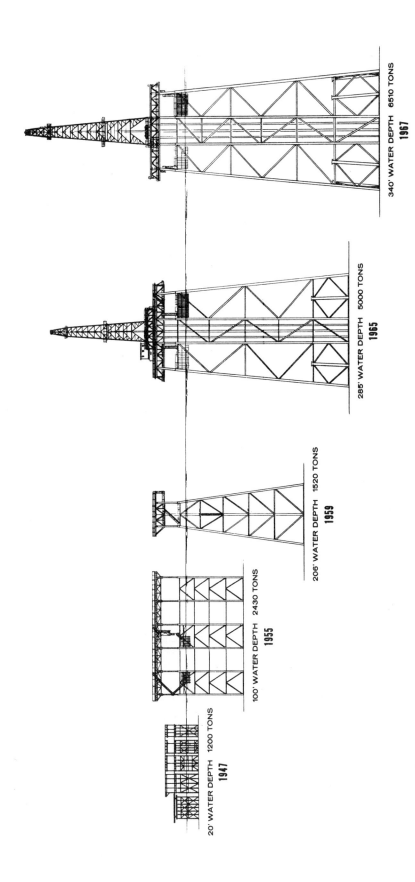

TWENTY YEARS OF PLATFORM DEVELOPMENT

20' WATER DEPTH 1200 TONS **1947**

100' WATER DEPTH 2430 TONS **1955**

206' WATER DEPTH 1520 TONS **1959**

285' WATER DEPTH 5000 TONS **1965**

340' WATER DEPTH 6510 TONS **1967**

Proceedings of OECON Offshore Exploration Conference — 1968, published by Offshore Exploration Conference, Palos Verdes Estates, California

"Offshore Structures Past, Present and Future Design Considerations" by G. C. Lee

- 163 -

or use tenders alongside for power, service or supplies and
ready evacuation in the event of storms. They are presently
limited to operations in depths of water less than about
350 feet.

Submersible Platforms - The first submersibles were barges
sunk in shallow water. This type of rig is mobile, having
a buoyant hull, which is used to float the rig on site, and
a fixed platform above. On location, the hull is then sunk
to the bottom. Submersibles presently have a depth capability
of up to about 175 feet, but this capability is limited by
the physical size of the individual platform, in length and
breadth dimensions as well as depth. Although mobile, this
type of platform is expensive to tow and is thus most
efficiently used within a limited area and in limited depths
of water.

Self-Elevating or Jack-Up Platforms - Self-elevating platforms
are towed to location floating on their platform-hull structure where pneumatic jacks, hydraulic jacks, or electric
rack-and pinion drives lower the legs to the sea floor and
then lift the hull above the sea surface to achieve the
required clearance. On some deep depth rigs, additional leg
sections are welded on as the rig is jacked up. Types in
which the footings for bottom support also serve as floata-
tion chambers and break the surface when afloat must be
rigged in a carefully controlled sequence. This critical
aspect of hydrostatic stability also afflicts the submersible platform type. Once rigged, these platforms are

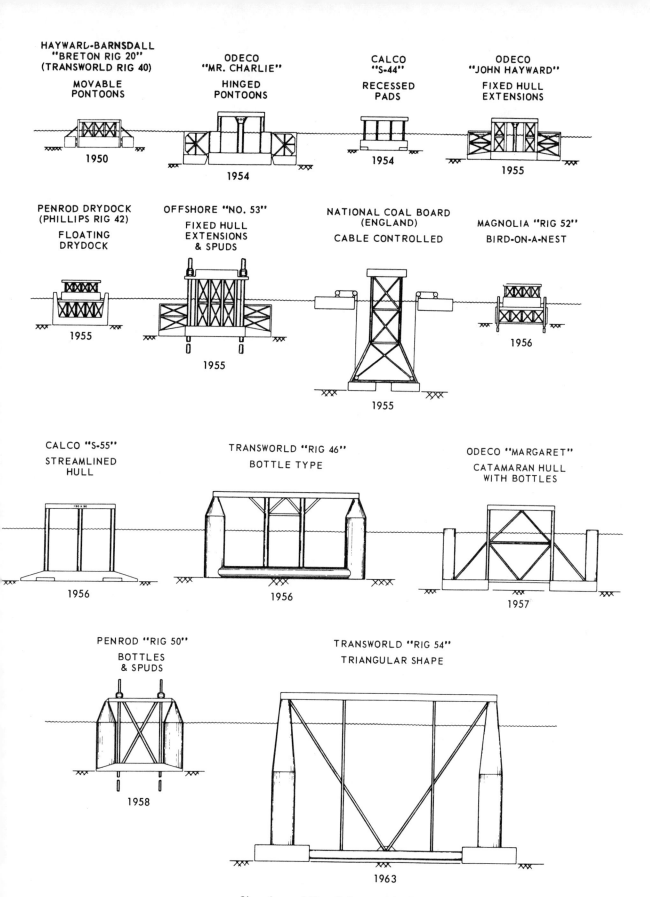

HAYWARD-BARNSDALL "BRETON RIG 20" (TRANSWORLD RIG 40)
MOVABLE PONTOONS
1950

ODECO "MR. CHARLIE"
HINGED PONTOONS
1954

CALCO "S-44"
RECESSED PADS
1954

ODECO "JOHN HAYWARD"
FIXED HULL EXTENSIONS
1955

PENROD DRYDOCK (PHILLIPS RIG 42)
FLOATING DRYDOCK
1955

OFFSHORE "NO. 53"
FIXED HULL EXTENSIONS & SPUDS
1955

NATIONAL COAL BOARD (ENGLAND)
CABLE CONTROLLED
1955

MAGNOLIA "RIG 52"
BIRD-ON-A-NEST
1956

CALCO "S-55"
STREAMLINED HULL
1956

TRANSWORLD "RIG 46"
BOTTLE TYPE
1956

ODECO "MARGARET"
CATAMARAN HULL WITH BOTTLES
1957

PENROD "RIG 50"
BOTTLES & SPUDS
1958

TRANSWORLD "RIG 54"
TRIANGULAR SHAPE
1963

Sketches of Key Submersible Rigs.

"Development of Offshore Drilling and Production Technology" by R. J. Howe
ASME Underwater Technology Division Conference - April 30-May 3, 1967

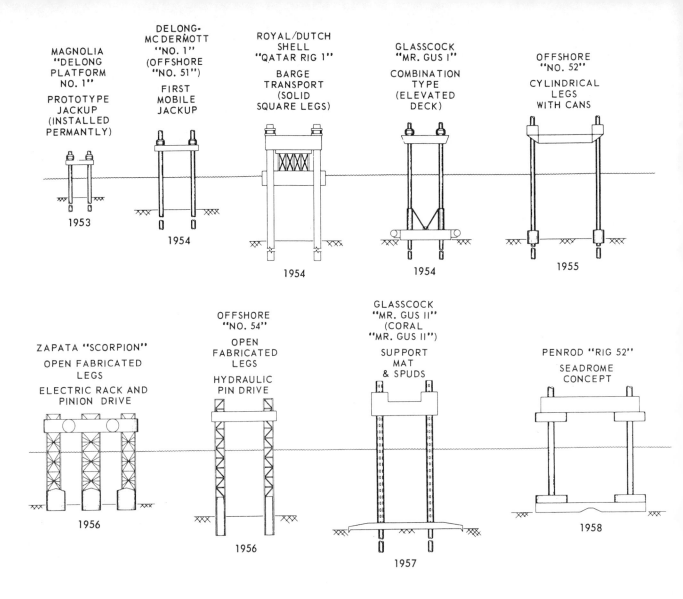

MAGNOLIA "DELONG PLATFORM NO. 1"

PROTOTYPE JACKUP (INSTALLED PERMANTLY)

1953

DELONG-MCDERMOTT "NO. 1" (OFFSHORE "NO. 51")

FIRST MOBILE JACKUP

1954

ROYAL/DUTCH SHELL "QATAR RIG 1"

BARGE TRANSPORT (SOLID SQUARE LEGS)

1954

GLASSCOCK "MR. GUS I"

COMBINATION TYPE (ELEVATED DECK)

1954

OFFSHORE "NO. 52"

CYLINDRICAL LEGS WITH CANS

1955

ZAPATA "SCORPION"

OPEN FABRICATED LEGS

ELECTRIC RACK AND PINION DRIVE

1956

OFFSHORE "NO. 54"

OPEN FABRICATED LEGS

HYDRAULIC PIN DRIVE

1956

GLASSCOCK "MR. GUS II" (CORAL "MR. GUS II")

SUPPORT MAT & SPUDS

1957

PENROD "RIG 52"

SEADROME CONCEPT

1958

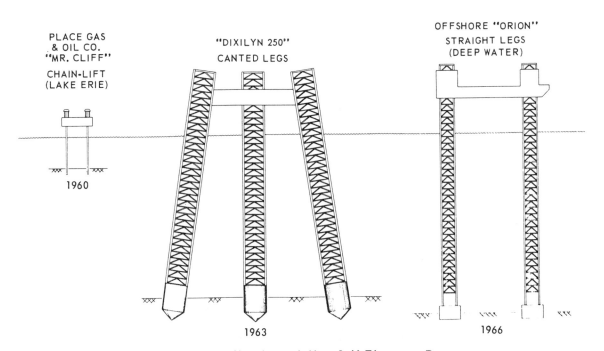

PLACE GAS & OIL CO. "MR. CLIFF"

CHAIN-LIFT (LAKE ERIE)

1960

"DIXILYN 250" CANTED LEGS

1963

OFFSHORE "ORION" STRAIGHT LEGS (DEEP WATER)

1966

Sketches of Key Self-Elevating Rigs.

"Development of Offshore Drilling and Production Technology" by R. J. Howe
ASME Underwater Technology Division Conference - April 30-May 3, 1967

highly stable in position and can continue operations in high seas. Footing areas must reflect the nature of the bottom support and in uncompacted sediments may result in a large mattress structure. Jack-up rigs are presently capable of operating at depths of up to 300 feet on legs as' much as 460 feet long.

FLOATING PLATFORMS

<u>Semi-Submersible Platforms</u> - Semi-submersible platforms can sometimes be used in relatively shallow water as bottom (submersible) units or in deep water as floating units. Semi-submersible rigs are so designed that when used as floating platforms the primary buoyancy is well below the water surface so as to be relatively unaffected by the most violent action of the surface waves. This causes a de-coupling of the platform motion from the surface motion and consequently semi-submersibles are much more stable than conventional ship hulls. This allows them to operate when conventional surface craft will have had to suspend operations. However, semi-submersibles are difficult to tow and they are much more expensive to build, transport, and insure. Being unwieldly to position, they are usually moored in position with an elaborate system of anchors, chain and tensioning devices. With dependence on bottom mooring, their depth capability is limited to shallower depths (up to about 1500 feet) than platforms relying upon dynamic positioning.

- 167 -

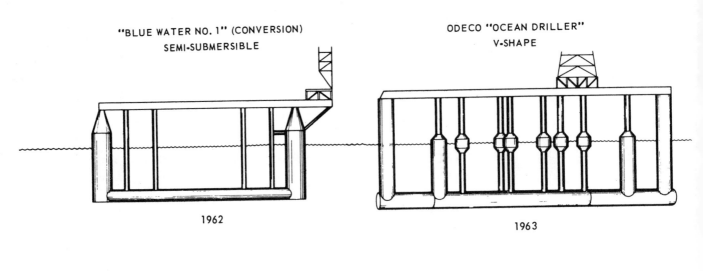

"BLUE WATER NO. 1" (CONVERSION)
SEMI-SUBMERSIBLE

1962

ODECO "OCEAN DRILLER"
V-SHAPE

1963

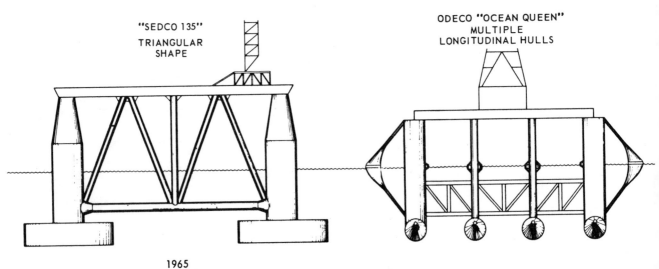

"SEDCO 135"
TRIANGULAR
SHAPE

1965

ODECO "OCEAN QUEEN"
MULTIPLE
LONGITUDINAL HULLS

Sketches of Key Semisubmersible Rigs.

"Development of Offshore Drilling and Production Technology" by R. J. Howe
ASME Underwater Technology Division Conference - April 30-May 3, 1967

Ship-type Platforms - Ship-type platforms, catamarans, or single hulls, are designed mainly for operation in deep water. Rigs of this type have been used for production drilling in depths of 600 feet and more. Sample bottom cores have now been obtained from 16,316 feet of water and 2,759 feet below the ocean bottom. The minimum depth of operation is determined by the positioning stability of the platform. For drilling this must probably be limited to a displacement between the two ends of the drill string of about three degrees from the vertical. The limit on horizontal excursion which this represents is probably more easily met in deep water than in shallow, despite the more positive positioning control presumably inherent in mooring versus dynamic positioning, as required in the great deeps. Ship-type platforms have been either towed or self-propelled. They are generally much less expensive than other types of floating rigs and insurance for them is less expensive. However, they are somewhat more limited than semi-submersibles as to the seas in which they may work.

Oil production is currently being carried on in water depths of up to 400 feet. Contemplating future possibilities at much deeper sites is likely to lead to submerged, pressure vessel platforms tethered to or supported by the ocean bottom and perhaps some distance above it.

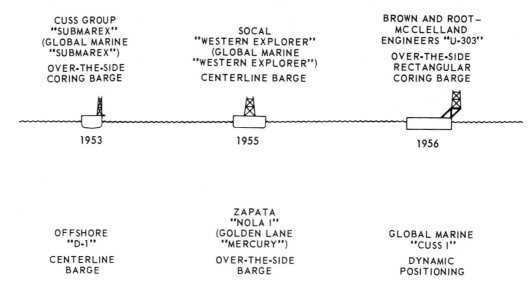

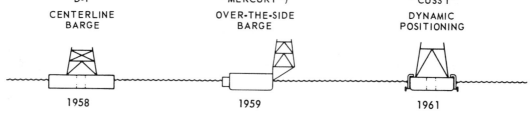

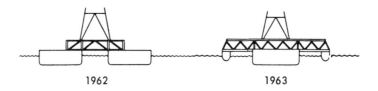

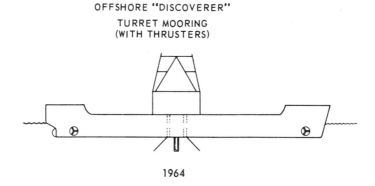

Sketches of Key Ship-Type Rigs.

"Development of Offshore Drilling and Production Technology" by R. J. Howe
ASME Underwater Technology Division Conference - April 30-May 3, 1967

Space Frame Analysis

The space frame is an especially important structural type in the field of ocean engineering. The use of such structures is particularly noteworthy in the case of bottom supported and many floating offshore work platforms, many of which are intricate and sophisticated examples of space frames.

Three dimensional framed structures are usually referred to as either "space trusses" or "space frames". In a truss all joints are assumed to be of the hinged type so that the members of the truss carry only axial loads. Therefore the joints of a truss each have three degrees of freedom, namely translational motion in the three coordinate directions. A space frame on the other hand has joints that can provide rotational restraint to the ends of the members, and thus these members carry both axial and bending loads. In this case each space frame joint has six degrees of freedom, corresponding to translation along and rotation about the three coordinate axes. In addition to the space truss and space frame, two dimensional networks are also used, called "plane trusses" and "plane frames". These are structures

all of whose members, loads, and deflections lie in one plane.
In the early, gross estimating stages of design, when
truss members are being sized by iterative methods, it
is often simpler and thus more efficient to handle the space
frame as an assemblage of plane frames, whenever possible.

The theory for the elastic analysis of both plane
and space frames and trusses has been well understood for
many years. However, practical application of theory to
complex actual problems has been difficult because of the
general nature of the basic equations. Statically determinant
problems, those for which member forces can be determined
solely by consideration of the equations of statics, although
readily solved, either occur extremely infrequently or are
poor models of the actual structures. As a result, up to
very recent times, analyses of such statically indeterminate
structures were carried out using any one of the several
available iterative calculation techniques, employing successive
approximations, or staged solutions, all of which are both
very tedious and time consuming. This situation has now
changed. The modern high speed digital computer has been
found to be ideally suited for the mathematical analysis of
framed structures. Based on the general matrix formulation
of a finite displacement solution, a number of computer programs
are now available which are capable of determining internal
loads and stresses and structural deformations and deflections

for frames and trusses, both space and plane. Such solutions require as input certain structural characteristics of the specific frame in question, namely a description of its geometry, the member properties, material properties, joint conditions, and the nature and magnitudes of the external loads.

The advent of computer solutions has had the effect of shifting the emphasis within the area of space frame analysis. The speed of such programs and the ease with which solutions can be produced has now focused attention on the problem of determining more reliably and precisely the magnitude and nature of the external loads acting on the structure. Our ability to analyze structures for given static loads has far surpassed our capability for accurately estimating what these design loads are.

Furthermore, waves of increasing length are possible in the greater water depths now becoming of interest. These waves may set up low frequency vibrations in bottom supported structures which may approach their resonant frequencies. Thus, there is also added interest in computer analyses for conditions of dynamic loading.

OFFSHORE DRILLING AND PRODUCTION PLATFORMS

Design Case Histories

The first drilling for underwater oil is believed
to have taken place in California in about 1900, where wells
were drilled from piers extending from the beach. Stationary
offshore platform construction in the open waters of the
Gulf of Mexico began in 1947 with the installation of the
first steel template type structure in 20 ft. of water A
second platform in a water depth of 50 ft. soon followed and,
according to G. C. Lee,"a new industry was born", although
several timber structures of many piles had been constructed
close to shore between 1938 and 1941.

The early platforms were supported by numerous
piles closely spaced, owing to the limited size of the steel
members then being manufactured and the limited capacity and
boom length of the floating construction equipment available
at the time. These restrictions were lessened with time so
that by the mid-1950's average pile sizes were 30" in diameter
and a 250 ton derrick was available for handling large templates and deck sections.

As water depths increased, it was increasingly
evident that deck sizes should be kept as small as possible.
Whereas the early platforms were entirely self contained, it
thus became customary to place the heavy equipment, drilling
supplies and living quarters on an attendant vessel or

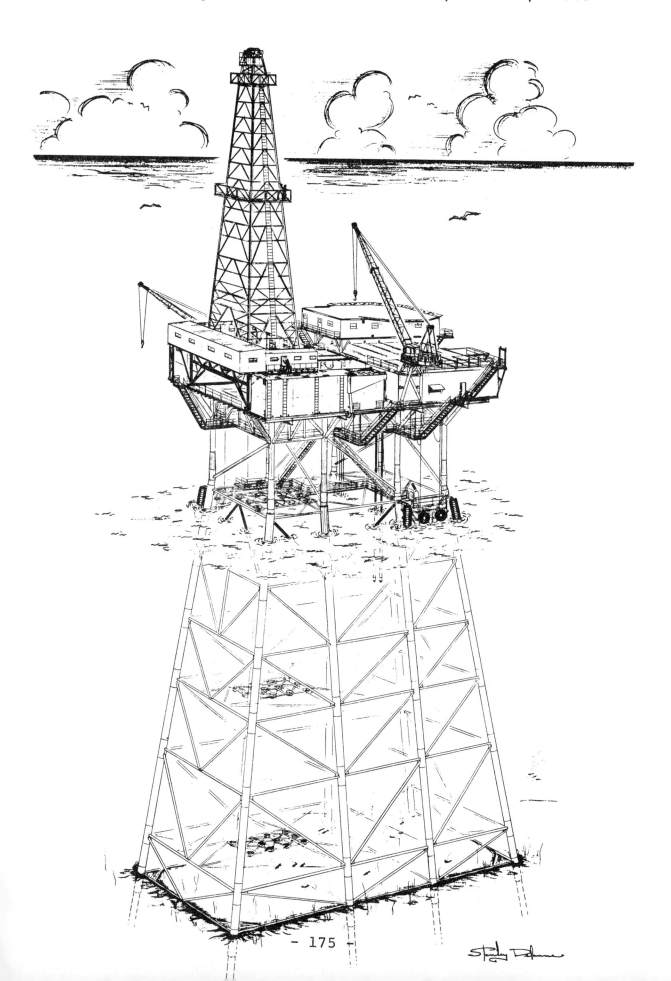

"The New Islands - The Offshore Platform", G. C. Lee,
El Colegio de Ingenieros Civiles de Mexico, June 24, 1966.

Fortune Magazine

Feb. 1965

"tender". On the debit side, it was necessary to suspend drilling operations earlier and move the tender away at the onset of bad weather. It follows that the subsequent trend was back to self-contained platforms, but more carefully arranged and with the minimum of supplies on hand to keep the deck areas low. Such deck sizes now average approximately 118' x 66'. In the Gulf of Mexico well over 2,000 platforms have been installed.

The first mobile offshore rig was a development from inland drill barges which appeared in South Louisiana in the early 1930's and were used in bays and marshes not over 10 feet deep. This limitation was imposed by the fact that if the deck were to become awash during submergence the rig would become unstable. First attempts to cope with this problem involved symmetrically arranged pontoon units sunk to provide bottom support before the central portion of the rig itself was placed on the sea floor. Additional hydro-static stability was given to later designs by increasing the diameter of the support columns penetrating the water surface. Most "bottles" of column stabilized units have tapered, conical tops for greater transparency to wave action on location.

Subsequently, such boot-strap devices largely passed out of favor owing to the vulnerability of their moving parts. A greater spread of columns followed and some platforms were designed to tilt until one end makes bottom contact before general submergence takes place by pivoting about the stabilized bottom corner.

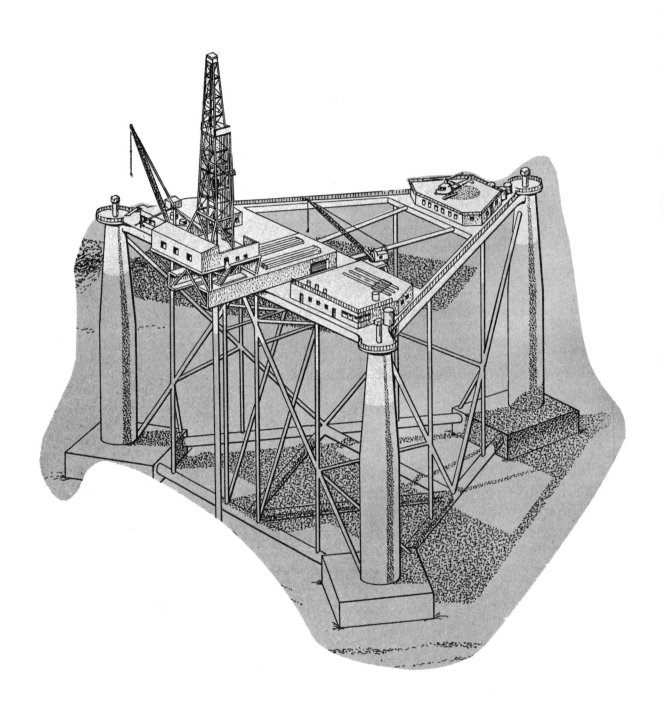

Fortune Magazine
Feb. 1965

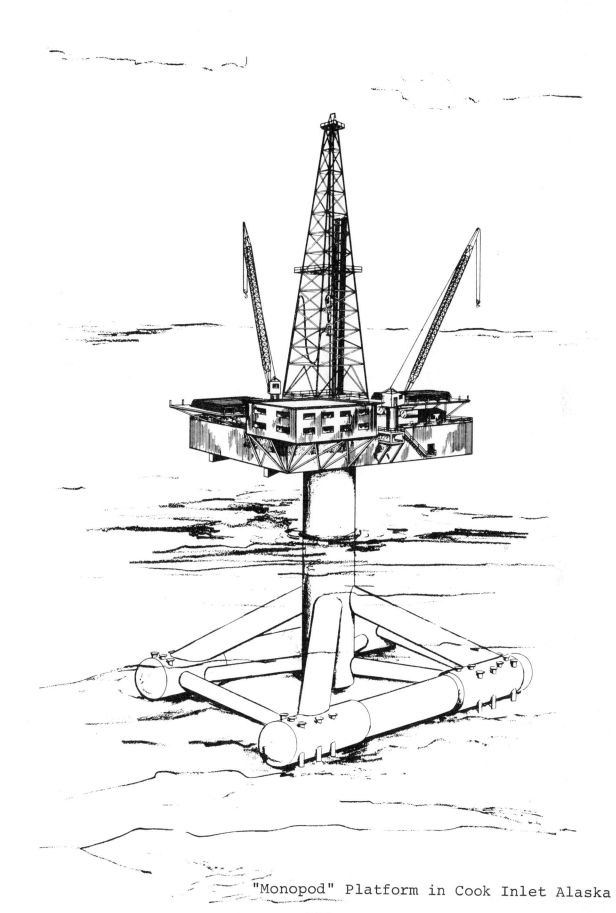

"Monopod" Platform in Cook Inlet Alaska

The number of jack-up rigs has increased rapidly since their appearance in the early 1950's until at present they constitute about 50% of the total, including ship types. A large factor contributing to their popularity is lower initial cost. An early difficulty was excessive leg penetration in soft soils. Aside from the structural overloads from differential settlement, there is increased probability of waves impacting against the platform; a primary source of concern. Large diameter "spud cans" for column footings eased the problem. As operating depths increase, bending stresses in cylindrical columns become critical prompting the use of open truss-type legs having greater strength without inducing larger wave forces. With especially soft, uninterrupted bottom conditions, a mat-type hull rigidly attached to the legs has been adopted. With over-the-side "work-over" rigs it is particularly beneficial. Dispensing with the mat has permitted canting the legs in some designs, thus minimizing leg bending moments and increasing stability, when extreme depth capabilities are the objective.

More attention is being paid to the hydrodynamics of submerged form under tow, now that drilling operations are being carried out in all parts of the world. Ship forms are most nearly ideal. These are self-propelled and suitable for higher speeds. To overcome trimming problems caused by large, variable hook loads, recent floating rigs (semi-submersibles or ships) have located the derrick at

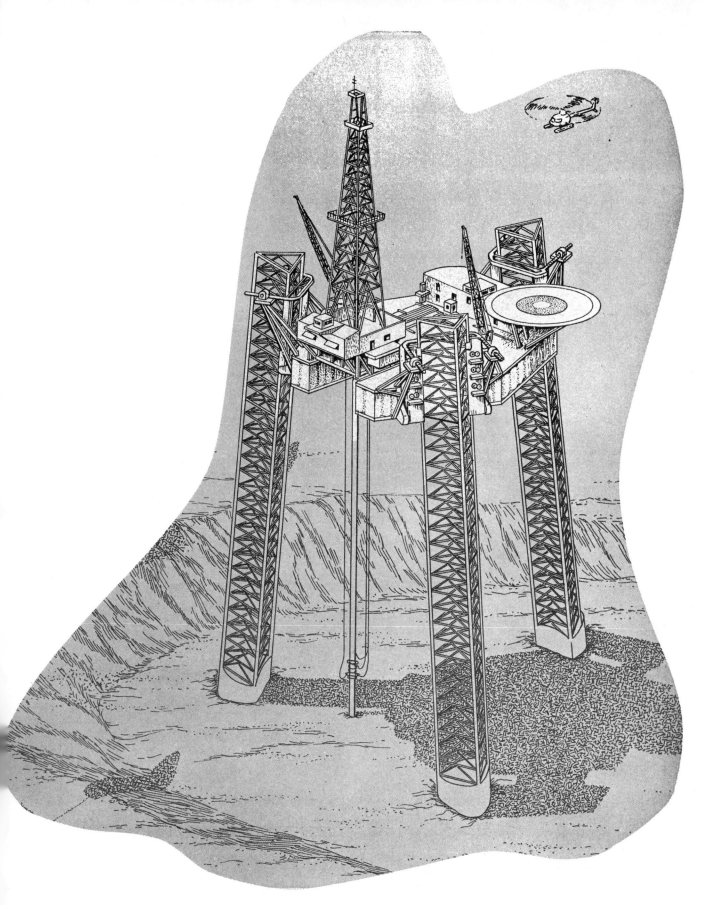

Fortune Magazine

Feb. 1965

- 181 -

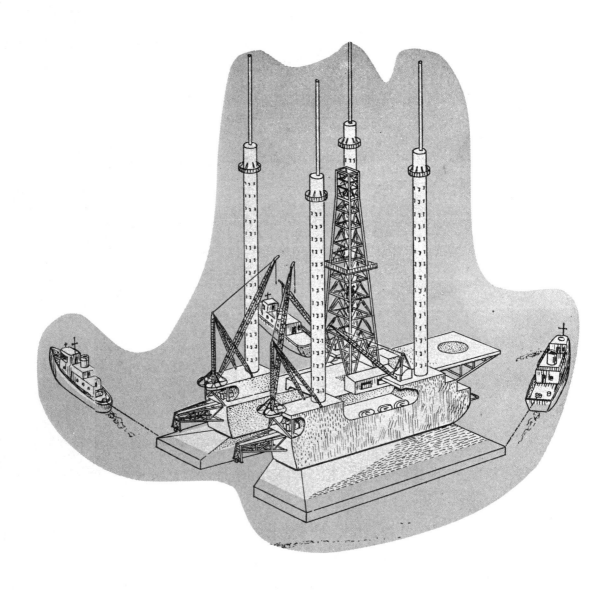

Fortune Magazine
Feb. 1965

- 182 -

the center of flotation, about amidships. This being near the center of pitch and roll oscillations in ships, it serves also to minimize the hook amplitude of vertical motion; the type of oscillation to which surface ships are most susceptible. The Catamaran ship forms now appearing are intended to reduce the amplitudes and accelerations of the basic motions themselves by conforming more nearly to the instantaneous wave slope, but improved seakindliness cannot be realized under all sea conditions.

According to Lee, the pattern emerging has most exploratory drilling being done by mobile rigs and development drilling from fixed platforms. Frequently, the fixed platform is set over the well which was previously drilled by a mobile rig. The remainder of the wells are then drilled from the platform. Separate platforms may be set in place for oil storage, quarters, compressors, separators, etc. (Of course, they have also been used for other purposes such as research or light stations.) For subsequent servicing, maintenance or reactivation of wells, mobile "work-over" rigs have been developed. Some are self-propelled.

An account of major failures in fixed platforms also is given by Lee as follows.

Hurricanes have caused the collapse of 22 platforms and severely damaged 10 others - to the extent that salvage was required. Two of these were temporary structures not intended to withstand storm; four were very early

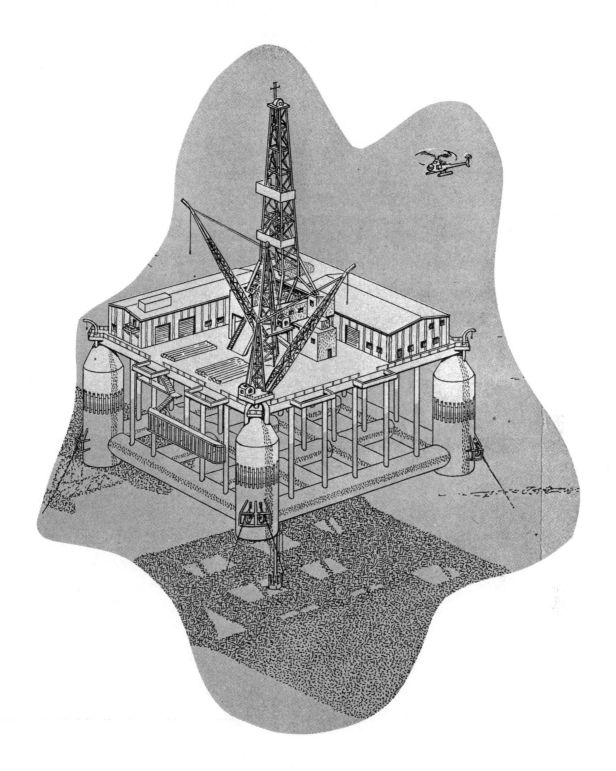

Fortune Magazine
Feb. 1965

Fortune Magazine

Feb. 1965

platforms which had been constructed before adequate design procedures or data was available. The remaining 26 platforms had been designed for a more up-to-date prediction of hurricane forces. However, 23 of these platforms had been designed to resist a 25-year probability storm. In an effort to reduce the initial cost, the owners had decided to accept a calculated risk in an effort to match the expected life of the structure with the possibility of disaster. Only three of the platforms which failed, all small well protectors, had been designed for the maximum storm conditions (as is true of practically all structures currently being installed in the Gulf of Mexico). The failure of two of these platforms was attributed to poor soil conditions; the other structure failed at "unreinforced joints" which were no longer considered adequate by the owner.

It appears that most of the fixed platform failures to date can be attributed to failures initiated at joint intersections in the underwater brace structure or template. Further, the initiation failures appear to be a combination of cyclic high stresses and notch sensitive non-homogeneous material. Some failures may have been initiated by pile-soil failures; however environmental conditions have prevented thorough on-the-spot evaluation to establish initial failure points.

Major mishaps to mobile rigs, up to July 1, 1968, have been tabulated by Howe as given below. Five accidents caused by blowouts have not been included as these infrequent, chance industrial hazards are unrelated to a particular rig design or its operation. Nor are they unique to the ocean environment.

Type of Rig

	On Location				
	Severe Storms	Normal Conditions	Moving on or off	Under Tow	Totals
Submersibles	2	1	1	- (2)	4 (6)
Jack-ups	1	2	6	6 (7)	15(16)
Barges	2	- (1)	-	1 (2)	3 (5)
Semisubmersibles	2	-	-	1	3
Totals	7	3 (4)	7	8(12)	25(30)

Earlier, Gaucher had given a similar compilation including cases of somewhat less severity. The cumulative totals are in parentheses.

The high incidence of accidents with rigs under tow (especially jack-ups) emphasizes first the importance of seamanship and of proper operating procedures in transit. The maiden voyage is a particularly dangerous period. Hydrostatic and hydrodynamic stability should be of the greatest concern. The effects of platform motions and their large accelerations should be anticipated.

Disengagement with the bottom and jacking up are almost as critical as the towing operation. Overloading and wracking during this transition period have caused leg failures ending in loss of the entire platform. In at least one instance, unexpected sinking of the spuds was involved. A heavy ground swell with legs barely clear of the bottom can be hazardous for them if subjected to impact with the bottom.

Based on the number of rig-years of service, by type, jack-up rigs have had the highest percentage of accidents. They are most vulnerable while moving on or off location or while under tow. During the latter, carrying the legs in some semi-retracted position to suit the prevailing conditions can be beneficial in adjusting metacentric height, mass moment of inertia and damping, all of which will affect stability, oscillatory motions, associated accelerations, and forces on the legs and their supporting foundations.

The submersibles seem to have compiled a more creditable record probably because they have no moving structural members and operate in shallower, more protected waters.

OFFSHORE DRILLING AND PRODUCTION PLATFORMS

Tubular Joints

Structural failures in tubular, ocean structures indicate that the tubular joints generally represent the most likely source of trouble. These failures were seldom by simple plastic deformation, but rather by one of several other more complicated failure modes.

In some cases brittle fracture occurred in a normally ductile material. In this mode a brittle cleavage occurs and propagates with little or no plastic flow, and at stresses which may be substantially below the yield point. Some of the factors which may initiate or propagate brittle behavior are low temperature, the presence of flaws or notches, multiaxial stress configurations, high strain rate, and residual stresses. Even subtilely different steels of the same class show markedly different susceptibility to this type of failure.

Fatigue failures occur when loads are cyclic in nature, as is inevitably the case in a sea. Small fatigue cracks tend to form in areas of stress concentration, as at joints, after a certain number of cycles and eventually grow and may lead to ultimate failure. The number of cycles to failure (the fatigue life) depends on the material, the magnitude of the load, and the severity of the stress concentration. For the range of load cycles encountered in surface platforms, the fatigue strength may be substantially less than the yield strength. In addition, surface roughness and corrosion have a detrimental effect on fatigue life. In ships, high stress-low cycle fatigue seems to be the part of the spectrum of most concern.

Weld failures have occurred as a result of poor workmanship (possibly due to field conditions) poor shop practices, or poor weld design or sequencing. Weld design must consider the choice of material, the welding sequence and ease of fabrication so that the welder is physically capable of forming an efficient joint. As a general rule, field welding should be avoided wherever possible so that conditions may be optimum.

If the material of the tubular trusses happens not to be homogeneous, but stratified radially, the layers of the tube may separate when the material is stressed radially in tension. This delamination then results from the material not developing its strength through the entire thickness and is attributable to a flaw in its manufacture. Seamless tubing seems less subject to this difficulty than welded pipe.

In a tubular joint the continuous member (usually the larger) is referred to as the chord and the other member as the branch. There are essentially four types of tubular joints:

Directly interwelded - The ends of the branches are welded directly onto the chord. This method is the simplest and is often the least expensive.

Gussetted - The ends of the branches are welded to a flat gusset plate which, in turn, is welded into the chord.

Combined Interwelding and Gusset Plates - A combination of the first two.

Cropped Ended - The ends of the branches are pressed

flat and then welded onto the chord. This type of joint is not well suited to large tubes and is structurally deficient. It should not normally be considered for offshore design.

With the choice of the proper material and welding techniques, the problem of joint design becomes one of minimizing stress concentrations, i.e. of obtaining the most uniform possible load transfer from the branch to the chord, and one of minimizing bending stress variations across the members. Complex joint problems are often not amenable to solution by the exact methods of applied mechanics.

One characteristic of tubular sections which makes for difficulty is their radial-flexibility as a consequence of small thickness/diameter ratios. Because of this, several methods are used to radially stiffen the chord. If the chord need not be hollow, an internal diaphragm plate may be used. When it must be hollow, the most efficient solution is to use thickened sections or high strength material in the chord in way of the joint. This method is probably more expensive and also leads to other problems where the thickened section meets the lighter section of the chord. Another solution is the use of ring stiffeners outside the chord.

Tubular joints are often designed with two branches intersecting the chord at one point; one in tension and the other in compression. When one is in tension and the other in compression, the function of the chord wall is to transfer the load from one branch to the other. It is most beneficial thus to have the branches intersect with each other as much as possible, so as to transfer a large portion of the load without involving the chord.

When three or more members meet in a joint, moments arising from the non-coincident intersection of their axes may not be insignificant. Present practice prefers to eliminate the problem.

The purpose of a gusset plate is to distribute the load over as wide an area as possible. In a multi-branch gusseted joint, the purpose of the gusset plate is to transfer the load through itself as much as possible. The trouble again arises where the relatively rigid gusset makes contact with the flexible tube. This is particularly a problem at those points where the gusset plate meets the branch where severe stress concentrations occur as a result of the sudden change of stiffness. This problem can be alleviated somewhat by tapering the plate where it first encounters the tube so as to make the change in restraint more gradual.

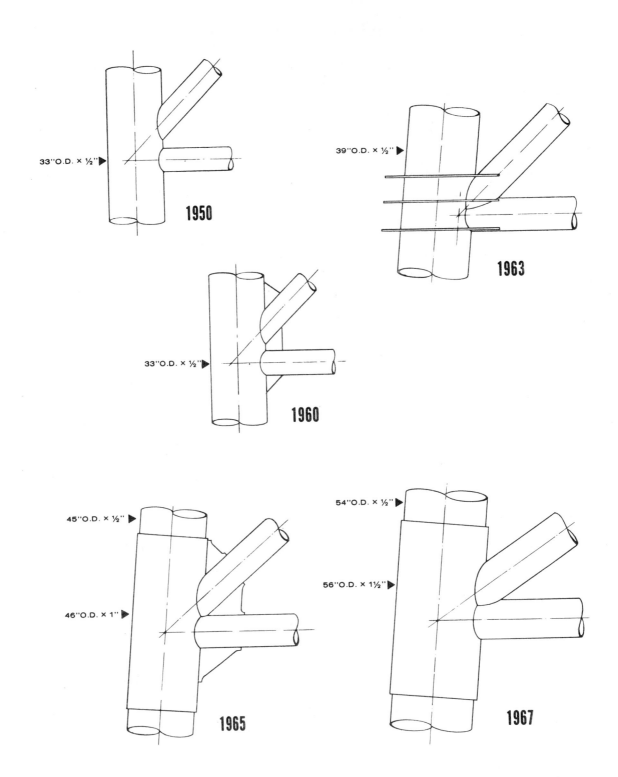

33"O.D. × ½" ▶ 1950

33"O.D. × ½" ▶ 1960

39"O.D. × ½" ▶ 1963

45"O.D. × ½" ▶
46"O.D. × 1" ▶ 1965

54"O.D. × ½" ▶
56"O.D. × 1½" ▶ 1967

"Offshore Structures, Past, Present, Future and Design Considerations" by G. C. Lee; Proceedings of OECON, 1968.

SUBMARINE VEHICLES

Design Requirements

Submarine vehicles are being called upon to perform an increasing variety of tasks. As a result, the list of functional requirements facing the designer itself continually grows longer. At the same time, the very hostile environment in which submarines operate in many ways limits the choice of the design parameters with which the designer works. As a result, a variety of vehicles, each to suit a specific purpose is inevitable. Although bottom supported or lowered vehicles should be included among them, the following comments apply specifically to floating vehicles having full freedom of motion in the hydrospace.

The principal purpose of a submarine being to operate submerged, the question of its maximum operating depth is of primary importance. Its choice can, and usually does, severely influence other design considerations such as the pressure hull configuration, the choice of material, cost, crew and payload capacity, speed, mobility, endurance, and others.

Already in submersible development, a number of means have been devised for providing mobility and maneuverability. The propulsion system may be completely self-contained or it may be absent from the vehicle itself, motion being obtained from towing by another vessel. The desire for a particular submerged velocity can be a critical requirement. Since the propulsion power varies roughly as the cube of the velocity, a moderate increase in velocity may require a substantial increase in power, resulting in increased propulsion equipment weight and space demands. This may have a derogatory effect in other areas, such as reduced payload, lower endurance, increased cost, etc. Furthermore, unline surface support vehicles, vertical maneuverability may also require a propulsion system.

Endurance usually refers to the length of time the vehicle may operate without surfacing (under normal operating conditions). Both men and machines are involved here in extraordinary degree. Energy storage must be sufficient not only to operate the propulsion, control and operational equipment, but also for the full life support system, all within severe restrictions. Provisions should also be made for fail safe operation in emergency situations.

The number of crew members, if the vehicle is manned, and the weight and/or volume of the desired payload is critical, bearing in mind the absolutely essential neutral buoyancy which the vessel must maintain in submerged operating condition.

If the vehicle is to be used at different geographical locations, and on call, consideration must be given as to how it is to be transported between sites. The necessity for rapid transportation by air, for example, may impose severe weight and size limitations. Set up time to become operational must also be minimized. Handling considerations can have a direct bearing on the topside configuration.

As hydrostatic and hydrodynamic stability must be satisfactory for the vessel in all operating conditions of ballast, and for both submerged and surface operating conditions, flexibility in the ballast system is mandatory. A deficiency of seaworthiness on the surface may seriously and unduly restrict the weather conditions in which the vehicle may be immersed or removed from the sea and severely limit its usefulness. In an emergency situation, such as caused by malfunction of equipment or entanglement, it may be permissible to accept some deficiency of stability, and even a negative metacentric height, so long as a positive buoyancy is assured.

Differential compressibility between sea water and the pressure hull will result in either loss or gain in buoyancy with increased depth of immersion and compensation will be needed.

Cost is much more likely to be a controlling factor in a commercial venture than in a military one. Even with small vehicles, this may lead to a desire for insurance coverage and, hence, to a need for classification by a body such as the American Bureau of Shipping. This implies adherence to a design and operational code of standards.

Hydrostatic pressure imposes extremely severe loads on the submarine pressure hull. The desire is obviously to maintain structural integrity above all, and thus any factor that influences the structural behavior must be given serious consideration; while at the same time, those which are design variables must be optimized for the least weight solution. Those factors of primary concern structurally include the pressure hull configuration (usually a stiffened cylindrical shell or a sphere or some combination of these), the workmanship, the circularity tolerances maintained, the compressibility of the structure, the weakening effects of penetrations and reinforcements, and the choice of material.

Submarines of the present, small non-military type are not completely self-sufficient and must depend upon a surface support mother ship for long range transport and servicing between dives for submarine and crew. Such surface operations, especially launching and retrieving the submarine are critical, from both the mechanical and sea-worthiness points of view. Surface transition considerations may require of the submersible a surface freeboard otherwise undesirable. A watertight retractable skirt extended through the topside escape hatch is one solution.

The leaking of watertight hull penetrations, as may be required for electric cable, is a source of much trouble. So too is the differential straining at access

hatches and viewing ports. Creep of viewing port materials
and spawling at the interfaces can result in loss of watertight integrity which may be greater at intermediate depths
than at the maximum.

To the well demonstrated attack and missile
carrying capabilities of military submarines have now been
added potentialities for oceanographic, geological, archeological, acoustic and fisheries research by means of
commerical vehicles already available. Others are being
designed and built for search and rescue missions, servicing ocean bottom work installations, salvage, pipe and
cable laying and for man-in-the-sea taxi service. There
are also recreational possibilities.

As cargo carriers, the submarine tanker is the
most attractive, for reasons both naval architectural and
economic, although its time has not yet come. Yet, with
the further development of nuclear power plants, a sufficient demand for the natural resources of the Arctic, or in
the face of a possible advantage to be derived from a fore-
shortened route beneath the North Polar ice cap, this conclusion may could soon be reversed.

SUBMARINE VEHICLES

Structural Types

Non-military submersible vehicles operating between the ocean surface and the ocean bottom have demonstrated rapidly increasing capabilities and sophistication in recent years. As a result, submarine vehicles are now actively performing a large number of work functions and are being seriously considered for others. Present functions lie in the general areas of scientific research, military weaponry, both offensive and defensive, as well as in specific areas such as search and rescue, taxi service, recreation, salvage operations, cable and pipe laying; to make a partial list.

For convenience, submersibles may be categorized as one of three types, manned untethered submersibles, manned tethered submersibles, or unmanned submersibles.

Manned Untethered Submersibles - This category consists of those vehicles which are most commonly called "submarines". Such manned vehicles are entirely self-contained, carrying with them their own supplies of power for mobility, for surfacing and submerging, and usually the means for life support. They employ no direct connection with the surface. Submarines vary in size from the very small recreational or research vehicles carrying a crew of one or two to the large military vessels, hundreds of feet in length and with crews in the hundreds. One self-contained submarine has demonstrated the depth capability to reach the deepest parts of the ocean. One class of small vehicles plays the role of a submarine "witches broom" to conserve the energies of men-in-the-sea and transport them to work sites.

Manned Tethered Submersibles - This category may be sub-divided into self-propelled submersibles and passive submersibles. The passive class includes such vessels as diving bells and the bathyspheres which are chambers lowered into the ocean on a support line or drawn down from below. The depth capability of those capsule systems with negative buoyancy may be limited by the ability of the lines to support their own paid out weight in addition to the submersible. Those systems with positive buoyancy have the problem of securing to the bottom. Power for equipment and communication is usually provided by addi-tional cables from the surface. Mobile tethered submersibles obtain their mobility in one of two ways. Some have their own propulsion and maneuvering motors and are connected to the surface by a cable whose only function is to provide power to these motors and to the vessels equipment. A tethered vehicle with surface support through a power cable has inherent advantages of power and endurance over one independently powered. But it also has limited depth and sea state operating capabilities. The towed resistance of the power cable adds a major component to the power require-ments. Other submarine types obtain their mobility by being towed by the surface vessel. In some of these the

submersibles have wing-like structures and maintain a
slight positive buoyancy, so that when moving submerged
they are in fact flying underwater as a kite.

Unmanned Submersibles - In this category there are both
tethered and untethered vehicles, the vast majority, however,
being tethered designs which are towed from a surface
vessel. Tethered unmanned submersibles have much in common
with the various types of the manned variety, but without
the necessity of providing life support systems. Consequently, there are some which are simply frameworks or enclosures
deisgned to support various instrument packages, and some of
which may be dragged along the bottom. There are other untethered, unmanned submersibles remotely controlled from the
surface. Mobility may be obtained through the use of the
usual propulsion motors or by buoyancy control with a winged
vehicle whereby successively increasing and decreasing the
buoyancy causes the submersible to rise and then glide down
in a cyclic manner.

The structures problem common to all submersibles,
in one form or another (whether it be the main hull or some
external component) is the provision of adequate pressure
bearing strength at or near the minimum cost in weight. For
pressure hulls with collapse depths greater than a few hundred feet, the provision of sufficient hydrostatic strength
will more than satisfy longitudinal strength requirements,
even when the vehicle is surfaced. Structural discontinuities, access and window openings and all hull penetrations

for cable leads, or the like, require special compensation, careful detail design and the best of workmanship. Cylindrical and spherical pressure hull configurations are almost universally used and great pains are necessary to strictly limit all out-of-roundness and residual stresses.

SUBMARINE VEHICLES

Configuration and Material Trade Off Studies

In the design of deep submergence pressure hulls,
as with almost all other structures, the final design inevitably represents a compromise between conflicting design
goals. The design process generally consists of studying the
various methods of satisfying the design requirements and then
partly trading off one goal against another so as hopefully
to reach the optimum design. In the case of deep submergence
pressure hulls one can consider the possible trade offs in
three basic areas of pressure hull design, choice of shell
configuration, shell construction, and the choice of material.
Probably the most common conflicts arise among cost, performance, and weight.

On a weight/strength ratio the sphere is the most
efficient configuration although cylinders are somewhat easier
to fabricate. However, there are other factors such as the
arrangement of internal volume, the external dimensions of the
vehicle or structure, the collapse depth, the vehicle hydrodynamics, as well as the problem of fabrication which can
influence the choice of shell configuration. As a result,
most small to moderate depth pressure hulls at present are made
up of combinations of spheres and cylinders. Thus the problem
becomes one of selecting that combination of spheres and
cylindrical shells, either interconnected or not, which best
meets the design requirements. Two of such combinations are

considered in the following figures. An appropriate criterion
of merit is the structural weight/buoyancy ratio, $W/_B$.
Excessive membrane stress is the assumed failure mode and not
instability on the assumption that this is the more likely
mode at all but small collapse depths. The basic solutions
assume shell intersections without reinforcement for stress
compensation. However, the second figure also shows results
with an allowance for internal reinforcement included. And in
one instance, external reinforcement is provided for in the
sense that the additional buoyancy of its displacement is taken
into account. The inference is that including consideration of
structural details may be (and in this case is) sufficient to
sway conclusions in the search for an optimum.

By far the simplest and least expensive form of shell
construction is the unstiffened shell. For cylinders within the
usual range of interest, it is also a type of construction of
less than optimum efficiency from the point of view of the
weight/strength ratio. Because of their instability behavior,
such shells are unable to take advantage of the high strength
levels available in high strength materials. This can be
remedied to a great extent by the use of rib stiffeners. Such
shells are capable of carrying higher load levels without becoming unstable and thus are able to take advantage of high
strength materials. The result is a shell with a lower
weight/strength ratio. The price paid for this improvement,
however, is increased construction cost and augmented stresses

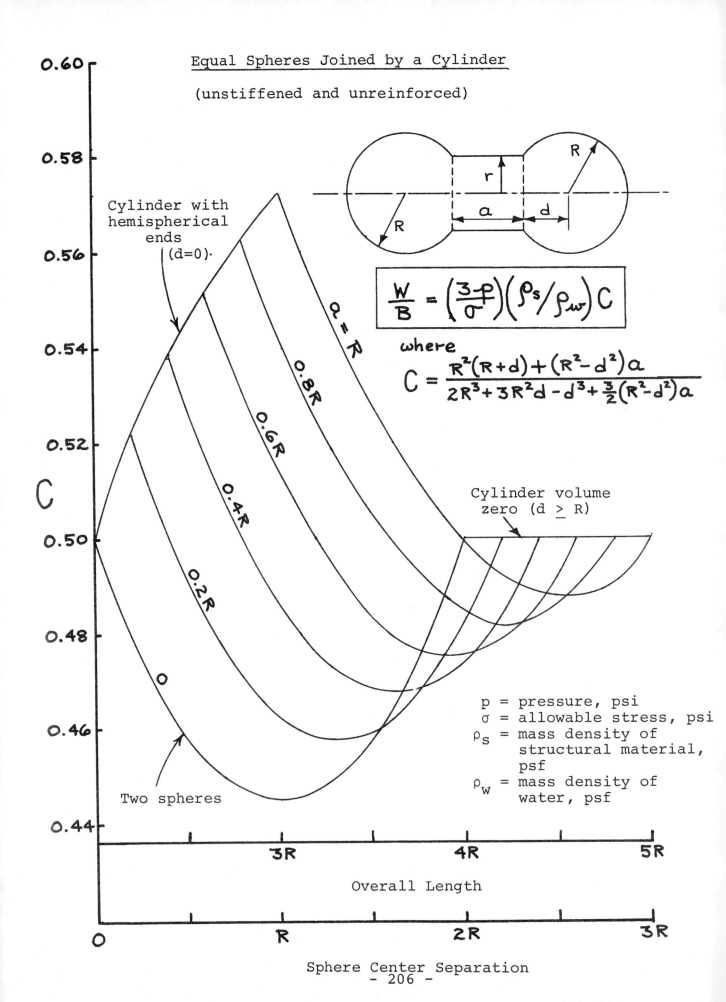

Equal Spheres Joined by a Cylinder

(unstiffened and unreinforced)

$$\frac{W}{B} = \left(\frac{3p}{\sigma}\right)\left(\rho_s / \rho_w\right) C$$

where

$$C = \frac{R^2(R+d) + (R^2-d^2)a}{2R^3 + 3R^2d - d^3 + \frac{3}{2}(R^2-d^2)a}$$

Cylinder with hemispherical ends (d=0)

Cylinder volume zero (d ≥ R)

$a = R$

0.8R

0.6R

0.4R

0.2R

0

Two spheres

p = pressure, psi
σ = allowable stress, psi
ρ_s = mass density of structural material, psf
ρ_w = mass density of water, psf

C

Overall Length

Sphere Center Separation

- 206 -

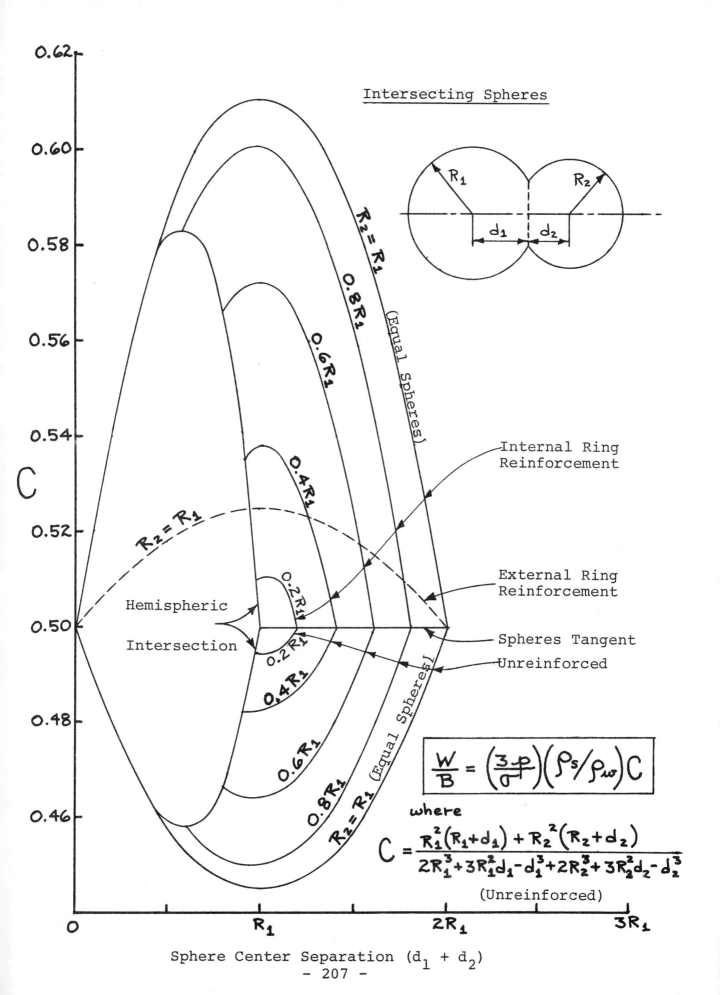

Intersecting Spheres

R_1 R_2 d_1 d_2

$R_2 = R_1$ (Equal Spheres)
0.8R₁
0.6R₁
0.4R₁
0.2R₁

$R_2 = R_1$

Hemispheric
Intersection

0.2R₁
0.4R₁
0.6R₁
0.8R₁
$R_2 = R_1$ (Equal Spheres)

Internal Ring
Reinforcement

External Ring
Reinforcement

Spheres Tangent
Unreinforced

$$\frac{W}{B} = \left(\frac{3-p}{\sigma}\right)\left(\rho_s/\rho_w\right)C$$

where

$$C = \frac{R_1^2(R_1+d_1) + R_2^2(R_2+d_2)}{2R_1^3 + 3R_1^2 d_1 - d_1^3 + 2R_2^3 + 3R_2^2 d_2 - d_2^3}$$

(Unreinforced)

C

0.62
0.60
0.58
0.56
0.54
0.52
0.50
0.48
0.46

O   R₁   2R₁   3R₁

Sphere Center Separation $(d_1 + d_2)$

near the ribs. Even more efficient shells are possible using rib cored sandwich shells and filled sandwich shells. These shell types usually consist of two high strength facings separated either by ribs or by a lightweight core material. The weight/strength ratios of such shells are lower than those of rib stiffened shells and, in the case of the filled sandwich shell, the nonuniform stress problem is eliminated. However, again the cost of such structures is at present somewhat higher than the others so that some compromise with cost must be worked out. Rib stiffened spheres are an anomalous form unlikely to be widely used.

In arriving at a choice of the best structural material to use, one is faced with another, similar type of problem. Those materials exhibiting very desirable high strength/weight ratios also tend to exhibit other undesirable features. These include high cost, problems with fabrication and/or joining, poor corrosion resistance, low fatigue life, and a variety of other unfortunate behavioral characteristics. This is true of metals, both common and exotic, as well as of other structural materials such as glass, ceramics, composite materials, and concrete. With some materials an additional problem is simply a lack of knowledge concerning their behavior. This is particularly true of a number of the non-metallic materials. However, this problem should become less as research on these materials continues.

For a given spherical shell thickness/radius ratio of 0.03, the third figure relates the weight/buoyancy ratios for different materials to the corresponding collapse depths. From such a figure, it is evident that the most desirable material at one collapse depth may not be so at another; quite apart from any consideration of secondary factors. Perhaps an even more revealing indication of a particular material's merits is given in the section on "Spherical Shells under External Hydrostatic Pressure."

The problem of trade offs in pressure hull design discussed here has been greatly simplified. These three areas of shell configuration, shell construction, and shell material are not completely independent variables as the above might lead one to believe. In fact, they are all somewhat interdependent in such a manner that one area cannot be considered without simultaneously considering the other two, and others. This simplified treatment is intended to illustrate the type of problems faced by designers of any type of ocean engineering structure, as well as those of deep submergence pressure hulls. With computer assistance, parametric studies and mathematical optimization techniques can be developed to help with such design decisions.

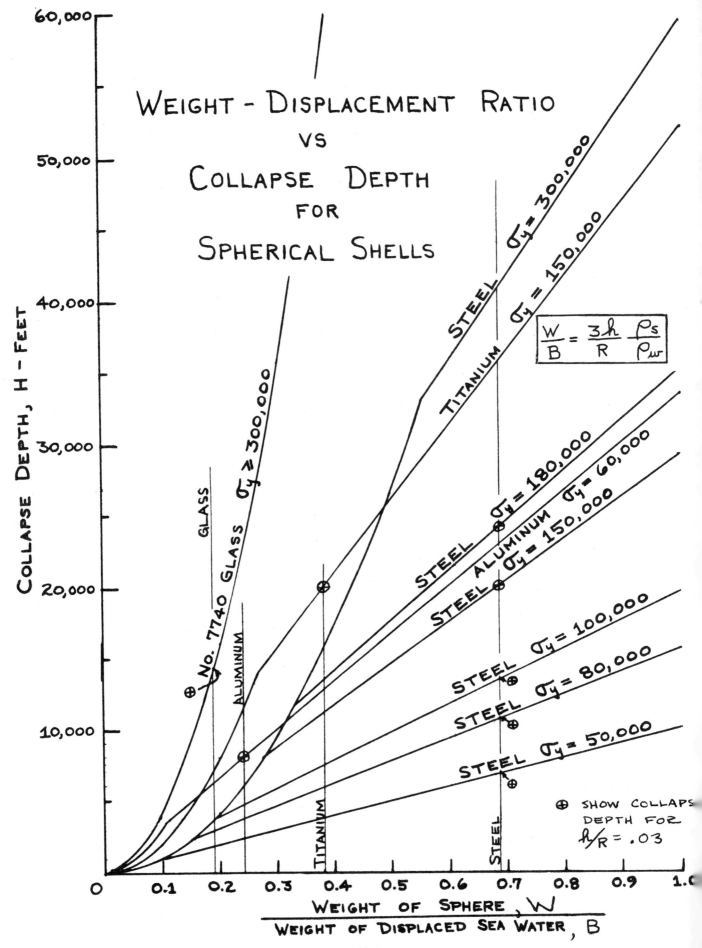

WEIGHT – DISPLACEMENT RATIO
VS
COLLAPSE DEPTH
FOR
SPHERICAL SHELLS

$$\frac{W}{B} = \frac{3h}{R} \frac{\rho_s}{\rho_w}$$

COLLAPSE DEPTH, H – FEET

STEEL $\sigma_y = 300,000$

TITANIUM $\sigma_y = 150,000$

GLASS $\sigma_y \geq 300,000$

STEEL $\sigma_y = 180,000$

ALUMINUM $\sigma_y = 60,000$

STEEL $\sigma_y = 150,000$

STEEL $\sigma_y = 100,000$

STEEL $\sigma_y = 80,000$

STEEL $\sigma_y = 50,000$

GLASS

No. 7740 GLASS

ALUMINUM

TITANIUM

STEEL

⊕ SHOW COLLAPSE
DEPTH FOR
$h/R = .03$

WEIGHT OF SPHERE, W
—————————————————
WEIGHT OF DISPLACED SEA WATER, B

SUBMARINE VEHICLES

Design Case Histories

The characteristics of some 35 non-military, manned submarine vehicles are available and provide some insight into the problem of submarine vehicle design. The structural features of particular interest are the configuration of the pressure hull and its material of construction. Thirty-three of the vehicles are at present operational or are under construction.

For purposes of design analysis, it is illuminating to relate the pressure hull characteristics to the operating depth of the vehicle. These operating depths fall within certain distinct regions as opposed to being more uniformly distributed and tend to cluster about certain specific values, reflecting where the majority of operations are to be carried out. Operations on the Continental Shelf, as usually defined, are limited to depths of less than about 600 feet or 200 meters (656 feet). The Continental Slopes fall off to about 12000 feet before meeting the ocean's basins. Only the very small percentage of bottom area included in the trenches lies below 20,000 feet and the greatest known depth is about 36000 feet.

Of the 31 vehicles in operation and whose operating depths are known, the pressure hulls of those 12 with operating depths up to 1,000 feet are constructed of medium steel of one type or another. The configuration of the pressure hulls is hardly uniform, consisting of spheres, stiffened cylinders, various unconventional shapes, and in one case an elliptical shell. This is not surprising. At these relatively shallow depths the hydrostatic pressure is not sufficiently severe to require using more structurally efficient hull shapes and/or materials. Consequently it is possible to sacrifice some of the structural efficiency to take advantage of the other desirable features offered by unconventional hull shapes and cheaper, more readily available, and more easily worked materials.

As depth capability increases, however, the effect of increasing hydrostatic pressure can be seen, both on configuration and material. The 10 vehicles in the operating range of 1200 to 2500 feet use one of the varieties of high strength steel, such as HY-80, HY-100, "T-1", or others. Their hull forms are the more conventional but more structurally efficient types, consisting either of stiffened cylindrical shells with hemispherical end

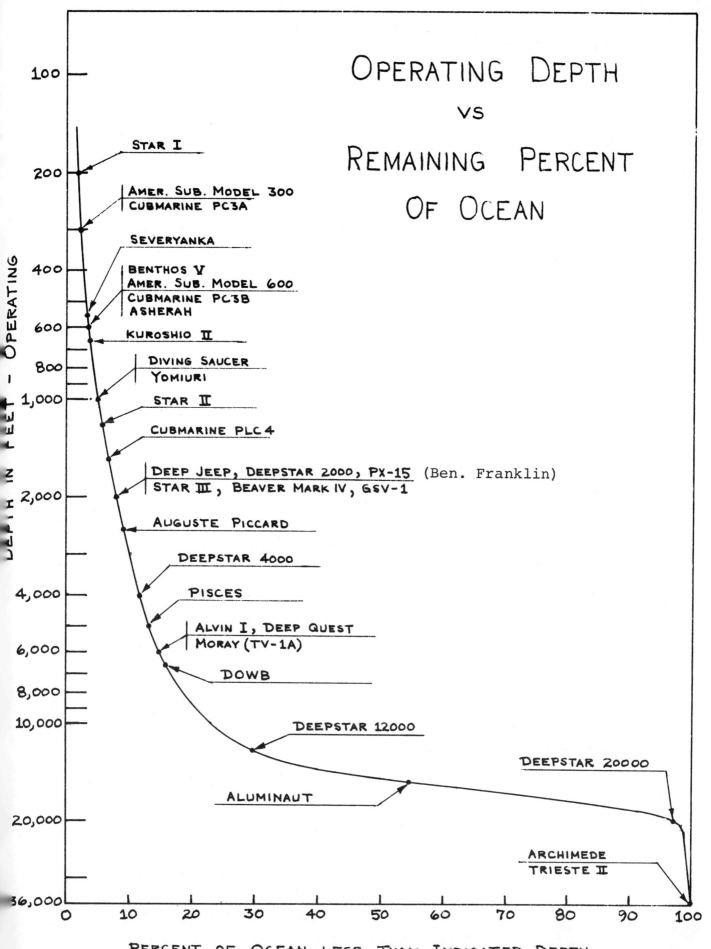

OPERATING DEPTH VS REMAINING PERCENT OF OCEAN

PERCENT OF OCEAN LESS THAN INDICATED DEPTH

caps, or they are spherical shells; the numbers of each being about evenly divided. One exception combines these two types, having two spheres connected by a cylindrical shell.

At operating depths in the range of 4000 to 6500 feet the trend toward the more structurally efficient hull forms continues. Out of 6 vehicles in this group, 4 have spherical shells, 1 has two separated spherical shells, and 1 has two intersecting spheres. High strength steels maintain their monopoly as construction material, but with one exception; the separated sphere hull uses a high strength aluminum alloy, A-356-T6.

For operating depths greater than 6500 feet, where lies more than 84% of the ocean bottom, few vehicles are to be found and few generalities can be drawn. Only 3 of the vehicles now in existance have a depth capability of greater than 6500 feet, one having a limit of 15000 feet and the other two of 36000 feet. These must be considered individually. The Aluminant is a relatively large vehicle. Its hull is close to 50 feet long and is made up of 11 forged cylindrical sections 8 feet in diameter with hemispherical heads. The sections are of 7079-T6 aluminum alloy 6.5 inches thick and are bolted together.

The two vehicles capable of 36,000 feet, Trieste II and Archimede, are fairly similar. Both use forged steel spheres for pressure hulls, relatively small in diameter (about 7 feet) but very thick (approximately 5 - 6 inches). Because of the negative buoyancy of the pressure hulls, resulting from the severe strength requirements, both require large floats for lift, making them also relatively larger than most other research submarine vehicles. "Fail-Safe" operation consists of jettisoning whatever solid ballast remains aboard at the time. In case of an emergency, the pressure capsule, if cast loose, could not by itself return personnel to the surface. They are in fact, a unique class of vessel and have limited performance capabilities, except for their operating depth.

A few general conclusions can be drawn from these data. The first is that except for very shallow depths, the more structurally efficient spheres, stiffened cylinders, and combinations of these two types have been used exclusively, for submarine vehicle pressure hulls. The second is that in spite of all the interest and development work on special materials, steel of one type of another, with very few exceptions, is the material of which pressure hulls are presently constructed. And third, there is a definite lack of practical vehicles with operating depths in the vicinity of 20,000 feet. A vehicle with such depth capability would be able to explore 99%

of the ocean floor as well as all the ocean deeps above it.

The two tables of structural characteristics are the basis for the second figure, dealing with overall structural efficiency. The figure is a rough and imperfect attempt to focus on design performance with a structural perspective.

For most present and forseeable purposes, the small, 2-3 man vehicle will probably predominate. Mobility and total system costs are primarily responsible for this limit. Within this size are many design variations from one vehicle to another; not merely of degree but of fundamental concept. Any attempt to order and rank the designs must be necessarily incomplete and deal with but one design aspect at a time. Nevertheless, such an attempt is implied in the figure where a ranking coordinate system has been established, based upon the hypothetical weight of a vehicle whose pressure hull is a pure sphere built of a steel with an allowable stress limit of 80,000 psi and an elastic modulus of 30×10^6 psi. The ratio of collapse depth to operating depth has been taken as 1.5 and the total weight of the vehicle as three times the weight of the spherical shell. These ratios must represent about the ultimate ideal in design and fabrication performance,

Table 1

Vehicle No.	Name	Owner/Operator	Dimensions (feet)	Operating Depth (feet)	Dry Weight - Diving Condition (Pounds)*
1	Aluminaut	Reynolds International, Inc./ Reynolds Submarine Service Corp.	51/8/14.25	15,000	162,000
2	Alvin I	Office of Naval Research/Woods Hole Oceanographic Institution	22/8/13	6,000	29,100
3	American Submarine Model 300	American Submarine Co.	13/4.2/4.75	300	2,200
4	American Submarine Model 600	American Submarine Co.	13/5.5/5.2	600	3,500
5	Archimede	French Navy	69/13/26.5	36,000	438,800
6	Auguste Piccard	Swiss Nat. Exposition Corp.	93.5/19.7/24	2,500	368,200
7	Benthos V	Lear Siegler, Inc.	11.3/6.1/6	600	4,200
8	Cubmarine PC3A	Perry Submarine Builders, Inc./ Ocean Systems, Inc.	19/3.5/6	300	4,790
9	Cubmarine PC3B	Perry Submarine Builders, Inc./ Ocean Systems, Inc.	22/3.5/6	600	6,350
10	Cubmarine PLC4	Perry Submarine Builders, Inc./ Ocean Systems, Inc.	24/4.5/8.7	1,500	17,900
11	Deep Jeep	Naval Ordnance Test Station, China Lake, California	10/8.5/8	2,000	8,000
12	Deep Quest	Lockheed Aircraft Co.	39.1/19/13.3	6,000	112,000
13	Deepstar 2000	Westinghouse Electric Corp.	20/7/8.5	2,000	14,000
14	Deepstar 4000	Westinghouse Electric Corp.	18/11.5/7	4,000	18,000
15	Diving Saucer	OFRS - J. Y. Cousteau Westinghouse Electric Corp.	10/10/5.3	1,000	7,000
16	Dolphin AG(SS)555	U. S. Navy	152/19/20	-----	1,568,000
17	DOWB	General Motors Defense Research Laboratory	16/8.5/6	6,500	14,720

*Including disposable ballast and/or flotation material

Data principally from "Undersea Vehicles for Oceanography," U. S. Interagency Committee on Oceanography.

Vehicle No.	Name	Owner/Operator	Dimensions (feet)	Operating Depth (feet)	Dry Weight-Diving Condition (Pounds)
18	Kuroshio II	Hokkaido Univ., Japan	36.7/7.15/10.4	650	25,800
19	Moray (TV-1A)	U.S. Naval Ordnance Test Station, China Lake, Calif.	33/5.3/5.3	6,000	22,400
20	NR-1	U.S. Navy/Special Projects Office	------	------	------
21	Pisces	International Hydrodynamics Co., Ltd., Vancouver, B.C.	16/11.5/9	5,000	14,560
22	Severyanka	USSR/All-Union Institute of Marine Fishery and Oceanography	240/22/15	550	------
23	Star I	Electric Boat Co. General Dynamics Corp.	10.1/6/5.8	200	2,750
24	Asherah	General Dynamics Corp./Univ. of Pennsylvania Museum	17/7.7/7.6	600	8,600
25	Star III	Electric Boat Co. General Dynamics Corp.	24.5/6.5/9	2,000	18,300
26	Trieste II	U.S. Navy/COMSUBPAC	67/15/18	36,000	492,800
27	Yomiuri	Mitsubishi/Yomiuri Shimbun Newspaper, Tokyo, Japan	48/8.2/9.2	1,000	------
28	Star II	Electric Boat Co. General Dynamics Corp.	17.76/5.33/7.62	1,200	10,000
29	Beaver Mark IV	North American Aviation	25/9.5/8.5	2,000	27,000
30	Benjamin Franklin (PX-15)	Grumman Aircraft Eng. Corp.	48.73/18.5/20	2,000	291,200
31	Deepstar 12000	Westinghouse Electric Corp.	18/11.5/7	12,000	18,000
32	Deepstar 20000	Westinghouse Electric Corp.	28/8/8	20,000	36,000
33	Deep Diver	Perry Submarine Builders, Inc./Ocean Systems, Inc.	22/-/-	1,350	18,500
34	Nai'a (Perry PC5C)	Pacific Submersibles, Inc.	22/4.6/-	1,200	11,470
35	SURV	British Navy	10.85/6.3/9.5	1,000	13,440

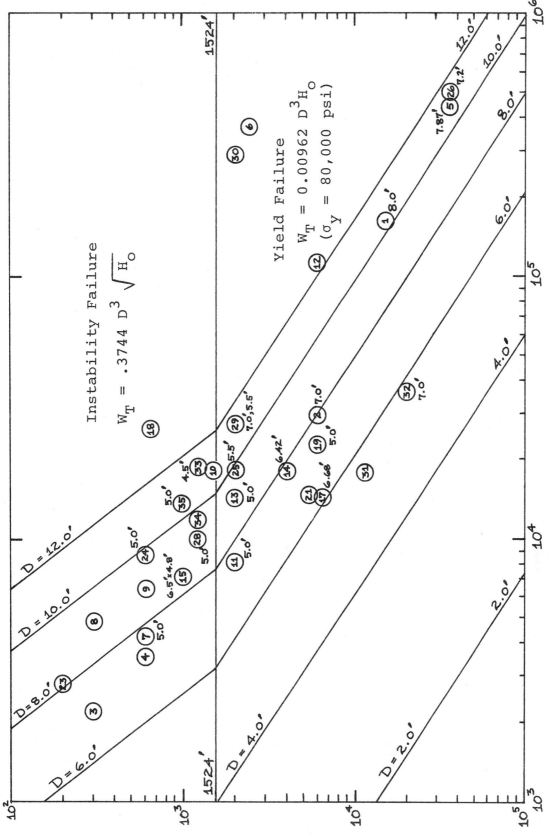

STRUCTURAL CHARACTERISTICS OF NON-MILITARY SUBMERSIBLES

Table 2

Vehicle No.	Name	Payload (Pounds)	Crew	Pressure Hull - Type & Diameter (feet)	RN_w	RN_D
1	Aluminaut	6,000	4 - 6	Aluminum alloy 7079-T6 integrally stiffened cylinder with hemispherical heads, 8.0 o.d., 7.0 i.d.	0.46	0.77
2	Alvin I	1,200	2	HY-100 steel sphere, 7.0 o.d., (1.33" thick).	0.68	0.88
3	American Submarine Model 300	450	2	A-36 steel, (0.375" thick)	----	----
4	American Submarine Model 600	750	2	A-36 steel, (0.5"thick)	----	----
5	Archimede	4,000	3	Ni-Cr-Mo forged steel sphere, 7.87 o.d., 6.9 i.d.	0.38	0.73
6	Auguste Piccard	20,000	40	Steel cylinder with hemispherical heads	----	----
7	Benthos V	400	2	Mild steel sphere, 5.0 o.d., (0.625" thick)	0.17	0.65
8	Cubmarine PC3A	750	2	A285 steel	----	----
9	Cubmarine PC3B	950	2	A212 steel, (0.5" thick)	----	----
10	Cubmarine PLC4	1,500	4	T-1 steel	----	----
11	Deep Jeep	200	2	HY-50 steel sphere, 5.0 o.d.	0.30	0.67
12	Deep Quest	3,400	4	Two intersecting maraging steel spheres	----	----
13	Deepstar 2000	1,000	3	HY-80 steel cylinder with hemispherical heads, 5.0 o.d., (0.75"thick)	0.17	0.56
14	Deepstar 4000	600	3	HY-80 steel sphere, 6.42 o.d., (1.2")	0.56	0.83
15	Diving Saucer	75	2	Mild steel ellipsoid, 6.5 major d., 4.8 minor d., (0.75" thick)	0.32	0.73
16	Dolphin AG(SS)555	-----	22	HY-80 cylinder with hemispherical heads	----	----
17	DOWB	1,021	2	HY-100 steel sphere, 6.68 i.d., (0.915" thick)	1.30	1.09
18	Kuroshio II	-----	4 - 6	Mild steel plate	----	----

Vehicle No.	Name	Payload (Pounds)	Crew	Pressure Hull – Type & Diameter (feet)	RN$_w$	RN$_D$
19	Moray (TV-1A)	200	2	Two aluminum A-356-T6 spheres, 5.0 o.d.	0.32	0.69
20	NR-1	-----	----	HY-80 steel	----	----
21	Pisces	1,500	2	Two Algoma 44 steel spheres	----	----
22	Severyanka	------	6 – 8	--------	----	----
23	Star I	200	1	A212 steel sphere, (0.375" thick)	----	----
24	Asherah	250	2	A212 steel sphere, 5.0 i.d., (0.625" thick)	----	----
25	Star III	1,000	2	HY-100 steel sphere, 5.5 i.d., (0.5" thick)	0.17	0.56
26	Trieste II	20,000	3	Ni-Cr-Mo forged steel sphere, 7.0 o.d., (3.5" thick) or forged steel sphere, 7.2 o.d., (4.72" thick)	0.24	0.64
27	Yomiuri	------	6	High tensile strength steel cylinder	----	----
28	Star II	250	2	HY-80 steel sphere, 5.0 i.d., (0.625" thick)	0.14	0.54
29	Beaver Mark IV	2,000	4 – 5	Two HY-100 steel spheres connected by 2.08 d. tunnel, 7.0 d., (0.481" thick), 5.5 d., (0.387" thick)	0.24	0.64
30	Benjamin Franklin (PX-15)	11,200	6	HY-80 steel ring stiffened cylinder with hemispherical heads, (1.375" thick)	----	----
31	Deepstar 12000	------	3	--------	----	----
32	Deepstar 20000	1,000	3	Steel (180,000 psi) sphere, 7.0 o.d., (1.85" thick)	1.83	1.22
33	Deep Diver	------	4	T-1 steel ring stiffened cylinder, 4.5 d., (0.5" thick)	0.06	0.40
34	Nai'a (Perry PC5C)	1,000	3	A212 Grade B steel, (0.5" thick)	----	----
35	SURV	------	2	Steel cylinder, 5.0 d.	0.09	0.48

especially in these small sizes. Thus, in the region of the
yield mode of failure (H_o > 1524 ft) for example, there
results the relationship:

$$W_T = 0.00962 \, D^3 H_o \quad \text{(pounds)}$$

with D the sphere diameter, and H_o the operating depth
in feet.

The number on each contour corresponds to the
hypothetical sphere diameter and may be related to the
actual diameter of most of the vessels plotted.

In Table 2 the weight rank number, RN_w, is the
inverse ratio of the actual weight of the particular
vessel to the weight of the hypothetical vessel of equal
pressure sphere diameter. The diameter rank number, RN_D,
is the ratio of diameters; that of the actual pressure
hull to that of the hypothetical sphere, for two vehicles
of equal weight.

Obviously, with only total weights available for
the 31 vehicles shown, it is not possible to factor out
excessive component weights of machinery and equipment
which may thus distort the true structural efficiency.

The ratio of collapse depth to maximum operating
depth represents a structural factor of safety and, of

course, may be varied with age and deterioration. Also,
with new and untried materials or inexperience, the factor
may be held on the order of 2.0 or more. But with
increasing confidence, the highest standards of welding
and fabricating workmanship, preliminary structural
model tests, and strain and deflection analyses of the
completed vehicle, ratios of 1.5 are generally believed not
to involve undue risk. Despite the losses of "Thresher"
and "Scorpion", it can be said that no known submarine
disaster in modern times can be attributed to failure of
the pressure hull at depths less than the specified operating depth even at safety factor values of 1.5.

Demands for increased visibility with overlapp-
ing cones of vision for pilot and observer, the ability
to observe a camera's objective, and to see upward in
ascent, will require a number of viewports; sometimes
closely spaced. Compensation for the loss of structural
integrity is thus a design problem whose future is assured.
Handling by the support vessel is equally important as a
consideration in future designs. Furthermore, the handling method used may affect many of the design
characteristics.

SALVAGE AND EMPLACEMENT OPERATIONS

Methods and Means

Salvage, being the broad term it is; including the recovery of a vessel, its components or what it contains from the effects of stranding, capsizing, collision, foundering, breaking up or the breakdown of its propulsion or control systems, its structural aspects are many and almost inseparable from hydromechanics, soil mechanics and other disciplines. By nature, salvage often has more the character of a rigging operation than of a design project. Perhaps largely because it is done in an emergency atmosphere, it seems to have fostered no body of theory unique to itself and often consists of extemporaneous applications of practical methods employing general purpose gear. This is not to imply that professional salvage operations are not highly organized, but rather that time is short and conditions highly variable. The obvious truth is that almost the total of all engineering effort is applied in creating devices that perform without serious failure and in assuring that in their use they may avoid major mishap or natural peril. These then are the overall bounding conditions within which salvage operations necessarily have been carried out, up to the present. Yet the importance of salvage must be acknowledged and, incidentally, its status as a venerable form of ocean engineering.

The other obvious but significant fact is that although ocean salvage efforts may be initiated with a vehicle's loss of power or control, the ultimate calamity involves loss of buoyancy which must be restored above all else, unless the final objective is merely to clear the obstruction.

In a stranding this may entail pulling off the reef with beach gear of rope, blocks and anchors embedded offshore. Jacking to ease the weight on the bottom may be used to assist.

After a capsizing in shallow water, a cofferdam from hull openings to the surface may effectively isolate the interior so that the water may be pumped out and the vessel refloated. For all the greater difficulty of control, compressed air may be resorted to for displacing the water. In the event of submerged openings to be closed, structural patches may have to be fitted.

Additional lift by external means is likely to be needed, whether lifting directly to the surface or only for transit to shallower water to obtain a new purchase. Now lines from padeyes just previously attached for the purpose, or from under the lift, to a surface barge or crane (or to pontoon strings) are required and a concentration of dynamic forces over a small area of structure again results.

In all these eventualities, structural loadings are imposed on the salvaged vessel which are quite foreign to those assumed in its design. Shoring-up or other measures for temporarily and locally strengthening the structure most severely affected are customarily a part of salvage procedures. Additional precautions must be taken when compressed air is used, to allow for the reducing external hydrostatic pressure during ascent. "Foamed-in" plastics which harden in place after being forced into a void space under pressure are a recent and happier solution so long as the difficulty with their ultimate removal is acceptable.

Questions of adequate longitudinal strength may arise, especially if the salvaged vehicle with its unusual weight and support distributions is required to travel the open sea, and in a holed condition. Further computational complexities in the stress analysis are introduced by the non-symmetry of the damaged cross section. The dangers are real. Vessels with temporary repairs to fortify their longitudinal strength have failed to make port even when the repairs were made by reputable repair organizations.

Special purpose surface craft designed and equipped for salvage work are not new, but the salvage systems of which they are a part seem destined to become more specialized and with them, the salvage vessels also. The greater variety of vessels to be recovered and the greater frequency of need are but one aspect of the change. Integration of the system components into a more unified, homogeneous systems design concept, and more precise specification of each component's function is, then, another.

Emplacement operations, as with the Sealab habitat or an ocean bottom nuclear power source, for example, while seeming to be merely salvage operations in reverse, have unique aspects of their own. From the structural point of view, this is most likely to be felt in the tension of a more positive attachment to the ocean floor for guidance, and controlled lowering precisely to a predesignated spot and there securing the containment vessel fast.

All the implications in future needs are impossible to foresee, but in the launching and retrieval of a small submarine by its support craft (a sort of semi-salvage operation) there is clear evidence that the two must be matched and all parts of the system made compatible. The emergence of double hulled, catamarans as mother ships is indicative of the novel solutions which may then result, and with them, new structures for analysis and design such as the cellular bridging structure between the hulls with its subjection to twist and edge moments.

Improving search techniques and capabilities for locating and homing on lost objects, and recent rapid advances in diving and work capabilities, encourage the belief that salvage and emplacement operations will ultimately be carried out in even the deepest parts of the world's oceans.

OCEAN ENGINEERING STRUCTURES

Volume II

(Selected References)

by

J. Harvey Evans

and

John C. Adamchak

PREFACE

The reference materials grouped in the lists
which follow were developed in association with a subject
of instruction in "Ocean Engineering Structures" given at
M.I.T. beginning in 1967. Notes on the subject matter
covered in that subject are given in Volume I. Papers,
journals and periodicals cited in the following include
those received through December 1, 1968.

OCEAN ENGINEERING STRUCTURES

Selected References

Contents

Page

SELECTED REFERENCES

Background Material

A. <u>Deep Ocean Environment and Resources</u>

1. Kort, V. G., "Antartic Ocean", <u>Scientific American</u>, Sept. 1962, p. 113.

2. Besse, C. P. and LeBlanc, N. F., "Application of Oceanographic Data in Offshore Structural Design", AIME, Society of Petroleum Eng. Paper SPE 1419.

3. Gordienko, P. A., "Arctic Ocean", <u>Scientific American</u>, May 1961, p. 88.

4. Trask, P. D., "<u>Applied Sedimentation</u>, John Wiley, 1950.

5. Bruun, "Bottom Roughness, Rivers, Tidal Inlets, Ocean", <u>Journal of Ocean Technology</u>, Vol. 1, No. 2, 1967, p. 1.

6. Taylor, D. M., "Billion-Dollar Scallop Find?", <u>Ocean Industry</u>, Dec. 1967, Vol. 2, No. 12, p. 20.

7. Bretschneider, C. L., "Calculating Storm Surge Criteria for the Continental Shelf", <u>Ocean Industry</u>, Dec. 1967, Vol. 2, No. 12, p. 42.

8. Glenn, A. H., "Comments on the Present State of the Sciences of Meteorology and Oceanography as Applied to Operational Forecasting and Hindcasting for Offshore Petroleum Operations", API, DPP. 1954.

9. Assur, A., "Composition of Sea Ice and its Tensile Strength", U. S. Army Cold Regions Research and Engineering Lab. Research Report No. 44, Dec. 1960.

10. Knauss, J. A., "Cromwell Current", <u>Scientific American</u>, April 1961, p. 105.

11. Bretschneider, C. L., "Decay of Wind Generated Waves to Ocean Swell by Significant Wave Method";
          <u>Ocean Industry</u>, Vol. 3, No. 3, March 1968, p. 36;
          <u>Ocean Industry</u>, Vol. 3, No. 4, April 1968, p. 45;
          <u>Ocean Industry</u>, Vol. 3, No. 5, May 1968, p. 54;
          <u>Ocean Industry</u>, Vol. 3, No. 6, June 1968, p. 99
12. Calhoun, J. C., Jr., "Food from the Sea: A Systems Approach", <u>Ocean Industry</u>, Vol. 3, No. 7, July 1968, p. 65.

13. "Engineering Properties of Marine Sediments", <u>U. S. Government Research Reports</u>, Vol. 38, No. 23, p. 11, December 5, 1963.

14. McClelland, B., "Engineering Properties of Soils on the Continental Shelf of the Gulf of Mexico", Eighth Texas Conference on Soil Mechanics and Foundation Engineering, Sept. 14-15, 1956.

SELECTED REFERENCES

Background Material

A. Deep Ocean Environment and Resources

15. Bretschneider, C. L., "Estimating Wind-Driven Currents Over the Continental Shelf", Ocean Industry, June 1967, p. 45, July 1967, p. 31.

16. Wilde, P., "Estimates of Bottom Current Velocities from Grain Size Measurements for Sediments from the Monterey Deep-Sea Fan", Ocean Sci. and Ocean Eng., Vol. II, MTS 1965, p. 718.

17. Darbyshire, M. and Draper, L., "Forecasting Wind-Generated Sea Waves", Engineering, April 5, 1963, p. 482.

18. Weeks, L. G., "Gas, Oil and Sulfur Potentials of the Sea", Ocean Industry, Vol. 3, No. 6, June 1968, p. 43.

19. "Operators Fight Icy Tides in Alaska", Oil and Gas Journal, January 13, 1964, p. 52.

20. Broodhead, G. C., "Fishery Science and the Prediction of Commercial Fish Landings", Exploiting the Ocean, MTS, 1966, p. 177.

21. Isaacs, J. D., "Food from the Sea", Int'l. Science and Technology, April 1967, p. 61.

22. Watson, E. E., "Gulf Stream", Science and the Sea (1967), U. S. Naval Oceanographic Office, p. 76.

23. Hironaka, M. C. and Smith, R. J., "Foundation Investigation for a Deep Ocean Materials Testing Structure", Civil Eng. in the Oceans, ASCE, San Francisco, Sept. 6-8, 1967.

24. Freeman, J. C., "Fundamentals of Ocean Weather",
    Ocean Industry, Vol. 2, No. 2, February 1967, p. 24;
    Ocean Industry, Vol. 2, No. 3, March 1967, p. 46;
    Ocean Industry, Vol. 2, No. 4, April 1967, p. 57;
    Ocean Industry, Vol. 2, No. 5, May 1967, p. 74;
    Ocean Industry, Vol. 2, No. 6, June 1967, p. 62;
    Ocean Industry, Vol. 2. No. 7, July 1967, p. 67.

25. Emery, K. O., "Geological Methods for Locating Mineral Deposits on the Ocean Floor", Exploiting the Ocean, MTS 1966, p. 24.

26. Ewing, M. and Ewing, J., "Geology of the Gulf of Mexico", Exploiting the Ocean (Supplement), MTS 1966, p. 145.

27. Bullis, H. R. Jr. and Thompson, J. R., "Harvesting the Ocean in the Decade Ahead", Ocean Industry, Vol. 3, No. 6, June 1968, p. 52.

28. Wilson, B. W., "How Rough are Waters of the Gulf", Offshore, October 1958, p. 58.

SELECTED REFERENCES

Background Material

A. Deep Ocean Environment and Resources

29. Chapman, W. McL., "How Space Research can Help Develop Fisheries",
     Ocean Industry, May 1967, p. 43.

30. Riehl, H. and Parks, M. H., "Hurricane Formation in the Gulf of
     Mexico", API, Paper No. 926-8-A, New Orleans, March 1963.

31. Blumberg, "Hurricane Probability in the Gulf of Mexico",
     Proceedings of OECON, 1967, p. 682.

32. Mayborn, D., "Hurricane Watch", Drilling, August, 1964, p. 50.

33. "Hurricane Wave Statistics for the Gulf of Mexico", Tech. Memo.
     No. 98, U. S. Army Corps of Engineers, June 1957.

34. Wilson, B. W., "Hurricane Wave Statistics for the Gulf of Mexico",
     6th. Int'l. Conf. on Coastal Engineering, Dec. 1957, p. 68.

35. Blumberg, R., "Hurricane Winds, Waves and Currents Test Marine
     Pipe Line Design",
             Pipe Line Industry, June 1964, p. 42;
             Pipe Line Industry, July 1964, p. 70;
             Pipe Line Industry, August 1964, p. 34;
             Pipe Line Industry, September 1964;
             Pipe Line Industry, October 1964;
             Pipe Line Industry, November 1964, p. 85.

36. Robin, G. deQ., "Ice of the Antarctic", Scientific American,
     Sept. 1962, p. 132.

37. LaFond and LaFond, "Internal Thermal Structures in the Ocean",
     AIAA Journal of Hydronautics.

38. Inderbitzen, A. L., "Investigation of Submarine Slope Stability",
     Ocean Sci. and Ocean Eng., 1965, MTS, Vol. 2, p. 1309.

39. Goodknight, R. C. and Russell, T. L., "Investigation of the
     Statistics of Wave Heights", ASCE Journal of the Waterways and
     Harbors Div., Paper No. 3524, May 1963.

40. Felando, A., "Kind of Oceanographic Information of Direct Use to
     the Fishermen", Exploiting the Ocean, MTS, 1966, p. 336.

41. Burke, W. T., "Legal Aspects of Ocean Exploitation - Status and
     Outlook", Exploiting the Ocean, MTS, 1966, p. 1.

42. Snyder, D. G., "Marine Protein Concentrate", Exploiting the Ocean,
     MTS, 1966, p. 530.

SELECTED REFERENCES

Background Material

A. Deep Ocean Environment and Resources

43. Mero, "Minerals on the Ocean Floor", Scientific American, December 1960, p. 64.

44. Shuleikin, V. V., "More Accurate Computation of Wind-Waves of Given Probability", Izv. Geophys. Ser., 1963/1, p. 156.

45. Fisk, H. N., "Nearsurface Sediments of the Continental Shelf off Louisiana", Eighth Texas Conference on Soil Mechanics and Foundation Engineering, Sept. 14-15, 1956.

46. Miller, A. R. et al, "Hot Brines and Recent Iron Deposits in Deeps of the Red Sea", U. S. Geol. Survey Report, open file series [No. 814] 1965.

47. Lyon, W., "Ocean and Sea-Ice Research in the Arctic Ocean via Submarine", Trans. of the New York Academy of Sciences, Ser. II, Vol. 23, No. 8, June 1961, p. 662.

48. Sette, O. E., "Ocean Environment and Fish Distribution and Abundance", Exploiting the Ocean, MTS, 1966, p. 309.

49. Murphy, R. C., "Oceanic Life of the Antarctic", Scientific American, Sept. 1962, p. 186.

50. Tilson, MacDonald, Hull and Craven, "Ocean's Resources and Working the Sea", Science and Technology, April 1967, pgs. 36, 38, 48, 50, 57.

51. "Oil under the Ice Floes", Surveyor, ABS, August 1967, p. 22.

52. Noble, V. E., "On the Decay of Wind-Driven currents", Ocean Sci. and Ocean Eng., Vol. I, MTS, 1965, p. 544.

53. Nelson, T. W. and Burk, C. A., "Petroleum Resources of the Continental Margins of the United States", Exploiting the Ocean, MTS, 1966, p. 116.

54. Bretschneider, C. L., "How to Calculate Storm Surges over the Continental Shelf",
    Ocean Industry, Vol. 2, No. 7, July 1967, p. 31;
    Ocean Industry, Vol. 2, No. 8, August 1967, p. 50;
    Ocean Industry, Vol. 3, No. 1, January 1968, p. 46.

55. Trask, P. D., Recent Marine Sediments.

56. Shepherd, F. P., "Recent Sediments, Northwest Gulf of Mexico", Tulsa, 1960.

57. von Arx, W. S., Introduction to Physical Oceanography, Addison-Wesley Publishing Company.

SELECTED REFERENCES

Background Material

A. Deep Ocean Environment and Resources

58. Bretschneider, C. L., "Maximum Sea State for the North Atlantic Hurricane Belt", Ocean Industry, Vol. 2, No. 9, Sept. 1967, p. 43.

59. Mero, J. L., "Review of Mineral Values on and under the Ocean Floor", Exploiting the Ocean, MTS, 1966, p. 61.

60. Isaacs, J. D., "Sampling Fish Populations and Acquisition of Fish of the Sea", ASME Underwater Tech. Conf., May 1967.

61. Assur, A. and Weeks, L. G., "Mechanical Properties of Sea Ice", U. S. Army Cold Regions Research and Engineering Laboratory, 1966.

62. Mero, J. L., "Mineral Resources of the Sea", American Elsevier Co.

63. McKelvey, V. E. and Chase, L., "Selecting Areas Favorable for Subsea Prospecting", Exploiting the Ocean, MTS, 1966, p. 44.

64. Blumberg, R., "Severe Hurricanes - A Look Ahead", Offshore, November 1965, p. 21.

65. Keller, G. H., "Shear Strength and Other Physical Properties of Sediments from Ocean Basins", Civil Eng. in the Oceans, ASCE, San Francisco, Sept. 6-8, 1967.

66. Miloy, L. F., "New Life-Saving Drugs are on the Way", Ocean Industry, Vol. 3, No. 6, June 1968, p. 74.

67. Kent, R. and Strange, R. R., "Some Aspects of Wave Forecasting on the Pacific Coast", Exploiting the Ocean, MTS, 1966, p. 211.

68. McClelland, B. and Focht, J. A., Jr., "Soil Mechanics as Applied to Mobile Drilling Structures, ASME Paper No. 55-PET-23.

69. "Ocean-bottom Minerals", Ocean Industry, Vol. 3, No. 6, June 1968, p. 61.

70. Shepard, F. P., Submarine Geology, Harper and Row, 2nd. Ed. 1963.

71. Sakou, T., "Surf on the Coral-Reefed Coast", Ocean Sci. and Ocean Eng., Vol. II, MTS, 1965, p. 700.

72. Fein and Freiberger, "Survey of the Literature on Shipboard Ice Formation", ASNE Journal, 1965, p. 849.

73. LaFond, E. C. and LaFond, K. G., "Temperature Structure in the Upper 240 Meters of the Sea", New Thrust Seaward, MTS, 1967, p. 23.

74. Webb, B., "Technology of Sea Diamond Mining", Ocean Sci. and Ocean Eng., MTS, 1965, p. 8.

SELECTED REFERENCES

Background Material

A. Deep Ocean Environment and Resources

75. Cuenca, R. M., "Utilization of the Difference in Temperature Between Deep and Surface Waters of the Sea", Ocean Sci. and Ocean Eng., MTS, 1965, Vol. 1, p. 188.

76. Terzaghi, K., "Varieties of Submarine Slope Failures", Eighth Texas Conf. on Soil Mechanics and Foundation Eng., Sept. 14-15, 1956.

77. Gaber, N. H. and Reynolds, D. F., Jr., "Ocean Engineering and Oceanography from the Businessman's Viewpoint", Ocean Sci. and Ocean Eng., Vol. I, MTS, 1965, p. 128.

78. "Wave Forecasting Relationships for the Gulf of Mexico", Tech. Memo. No. 84, Beach Erosion Board, Corps of Engineers, Dec. 1956.

79. Harrer, P., Blake, D., and Faughn, J., "Weather and Wave Forecasting for California Offshore Operations", API, Div. of Production, Paper No. 801-33-K, Los Angeles, May 16-17, 1957.

80. "Weather Averages and Sea States for Selected Offshore Areas",
    Ocean Industry, Vol. 2, No. 3, March 1967, p. 12;
    Vol. 2, No. 4, April 1967, p. 10;
    Vol. 2, No. 5, May 1967, p. 14;
    Vol. 2, No. 6, June 1967, p. 10;
    Vol. 2, No. 7, July 1967, p. 16
    Vol. 2, No. 8, August 1967, p. 6;
    Vol. 2, No. 9, September 1967, p. 18;
    Vol. 2, No. 10, October 1967, p. 12;
    Vol. 2, No. 11, November 1967, p. 8;
    Vol. 2, No. 12, December 1967, p. 6;
    Vol. 3, No. 1, January 1968, p. 14;
    Vol. 3, No. 2, February 1968, p. 18; etc.

81. Davis, J. A., "What Drillers can expect from the North Sea Weather", Oil and Gas International, July 1964, p. 30.

82. MacDonald, G.J.F., "What's in the Ocean?", Int'l. Sci. and Technology, April 1967, p. 38.

83. Freeman, J. C. and Gade, H., "Winds, Tides, Waves and Wave Forces in Hurricane Flossie of 1956", API Paper No. 926-2-G, Shreveport, March 57.

84. Shepard, F. P., Earth Beneath the Sea, Johns Hopkins Press.

85. Cruickshank, M., "Ocean Exploitation", Undersea Technology, Oct. 1963, p. 16.

86. Wiegel, R. L., Oceanographical Engineering, Prentice-Hall, Inc.

SELECTED REFERENCES

Background Material

A. <u>Deep Ocean Environment and Resources</u>

87. James, R. W., "Ocean Thermal Structure Forecasting", U. S. Naval Oceanographic Office SP-105 ASWEPS Manual, Volume 5.

88. Lermond, J. W., "Peru Current", <u>Science and the Sea (1967)</u>, U. S. Naval Oceanographic Office, p. 61.

89. Pierson, W. J., Neumann, G. and James, R. W., "Practical Methods for Observing and Forecasting Ocean Waves", U. S. Hydrographic Office Publ. No. 603.

90. Neumann, G. and Pierson, W. Jr., <u>Principles of Physical Oceanography</u>, Prentice-Hall, Inc.

91. Bretschneider, C. L., "Significant Waves and Wave Spectrum", <u>Ocean Industry</u>, Vol. 3, No. 2, Feb. 1968, p. 40.

92. Martin, O. L., "Titantic – 50 Years Later", <u>Science and the Sea (1967)</u>, U. S. Naval Oceanographic Office, p. 35.

93. Chanslor, J. W., "Treasure from the Sea", <u>Science and the Sea (1967)</u>, U. S. Naval Oceanographic Office, p. 9.

94. Chanslor, J. W., "Water Density and its Applications", <u>Science and the Sea (1967)</u>, U. S. Naval Oceanographic Office, p. 69.

95. Bretschneider, C. L., "Wave Forecasting", <u>Ocean Industry</u>, Vol. 2, No. 10, Oct. 1967, p. 53; Vol. 2, No. 11, Nov. 1967, p. 38.

96. Jordan, C. L., "What are Hurricanes and Where and When do they Occur?", <u>Ocean Industry</u>, Vol. 2, No. 8, August 1967, p. 16.

97. Browning, D. C., "Who has What Rights on the Continental Shelves", <u>Ocean Industry</u>, Vol. 3, No. 2, February 1968, p. 52.

98. Bowen, R. L., "Will the Pacific Island Areas Produce Oil?", <u>Ocean Industry</u>, Vol. 3, No. 6, June 1968, p. 77.

99. Mero, J. L., "Exploration for and Production of Solid Minerals from Offshore Area", <u>Proceedings of OECON</u>, 1966, p. 735.

100. Pierson, W. J., "Ocean Waves", <u>ASNE Journal</u>, Oct. 1964, p. 770 and <u>International Science and Technology</u>, June 1964.

101. Howe, R. J., "Petroleum Operations in the Sea – 1980 and Beyond", <u>Ocean Industry</u>, Vol. 3, No. 8, August 1968, p. 25.

102. Bascom, W. "Waves and Beaches", Educational Services Inc., 1964.

103. McKelvey, V. E., "Mineral Potential of the Submerged Parts of the U.S.", <u>Ocean Industry</u>, Vol. 3, No. 9, Sept. 1968, p. 37.

104. Overall, M. P., "Mining Phosphorite from the Sea", <u>Ocean Industry</u>, Part I, Vol. 3, No. 9, Sept. 1968, p. 44.

SELECTED REFERENCES

<u>Background Material</u>

A. <u>Deep Ocean Environment and Resources</u>

105. Monney, N. T., "Engineering Evaluation of Marine Sediments", <u>Proceedings of OECON</u>, 1968, p. 209.

106. Mero, J. L., "Exploration for and Production of Solid Minerals from the Offshore Area", <u>Proceedings of OECON</u>, 1966, p. 735.

107. Coene, G.T., "Future Engineering Possibilities in Recovery of Ocean Resources", <u>Ocean Industry</u>, Vol. 3, No. 11, Nov. 1968, p.53.

108. McKelvey, V. E., "Mineral Potential of the Submerged Parts of the U.S.", <u>Ocean Industry</u>, Vol. 3, No. 9, Sept. 1968, p. 37.

109. Overall, M. P., "Mining Phosphorite from the Sea",
<u>Ocean Industry</u>, Part I   Vol. 3, No. 9, Sept. 1968, p. 44
<u>Ocean Industry</u>, Part II, Vol. 3, No. 10, Oct. 1968, p. 60
<u>Ocean Industry</u>, Part III, Vol. 3, No. 11, Nov. 1968, p. 51

110. Brahtz, J. F. (Editor) <u>Ocean Engineering</u>, John Wiley and Sons, Inc.

111. Pierson, W. J., "Ocean Waves", <u>ASNE Journal</u>, Oct. 1964, p. 770 and <u>International Science and Technology</u>, June, 1964.

112. Bowen, R. L., "Parameters Controlling Offshore Concentrations of Non-Metallic Mineral Deposits", <u>Proceedings of OECON</u>, 1968, p. 775.

113. Howe, R. J., "Petroleum Operations in the Sea - 1980 and Beyond", <u>Ocean Industry</u>, Vol. 3, No. 8, August 1968, p. 25.

114. Shigley, C. M., "Seawater as Raw Material", <u>Ocean Industry</u>, Vol. 3, No. 11, Nov. 1968, p. 43.

115. "Updating Ocean Bottom Mineral Activity", <u>Ocean Industry</u>, Vol. 3, No. 11, Nov. 1968, p. 39.

116. Bascom, W., "Waves and Beaches", Anchor Books (S34) Doubleday and Co., Inc.

SELECTED REFERENCES

Background Material

B.   General Design Considerations

1.   Domiano, C. and Potthoff, "Arabian Oil Automates all the Way at Khafji", Petroleum Engineer, June 1964, p. 49.

2.   "Construction at Sea", Undersea Technology, Dec. 1963, p. 24.

3.   Talkington, H. R., "Deep Ocean Engineering", Civil Eng. in the Oceans, ASCE, San Francisco, Sept. 6-8, 1967.

4.   Rogers, L. C., "Deepwater-Positioning Problem Solved for Mohole Rig", Oil and Gas Journal, Sept. 21, 1964, p. 196.

5.   Rechtin, E. C., Steele, J. C. and Scales, R. E., "Engineering Problems Related to the Design of Offshore Mobile Platforms", SNAME, 1957, p. 633.

6.   Wilimovsky, "Engineering Aspects of the Parameters Regulating Fisheries", ASME Underwater Tech. Conf., May 1967.

7.   "Engineering for Inner Space", ASNE Journal 1965, p. 483.

8.   "Engineering for Marine Exploration", Undersea Technology, March, 1965, p. 29.

9.   Bullis, "Engineering Needs for Fishery Development", Exploiting the Ocean, MTS, June 1966, p. 342.

10. Craven and Searle, "Engineering of Sea Systems", Exploiting the Ocean, MTS, June 1966, p. 412.

11. Pritzlaff, "Functional Modules as an Approach to Underwater Vehicle Operation", ASNE Journal, 1963, p. 591.

12. Bretschneider, C. L., "Fundamentals of Ocean Engineering", Ocean Industry, Vol. 2, No. 6, June 1967, p. 45; Vol. 2, No. 7, July 1967, p. 31.

13. "Hardware Development Paces Undersea Activity", Undersea Technology, Jan. 1967, p. 53.

14. Rogers, L. C., "'Hilda' Kicks Operating Costs Up", Oil and Gas Journal, June 21, 1965, p. 154.

15. "Hurricane Betsy Swallows Zapota's Maverick Rig", Petroleum Engineer, Oct. 1965, p. 82.

SELECTED REFERENCES

Background Design

B. General Design Considerations

16. Heller and Abramson, "Hydroelasticity: A New Naval Science", ASNE Journal, 1959, p. 205.

17. Melson, "Implications of the Man-in-the-Sea Program", ASME Paper 66-MD-15.

18. "Journey into the Earth", Mechanical Engineering, July 1965, Vol. 87, No. 7, p. 27.

19. Wenzel and Helvey, "Manned Aspects of Deep Submersibles", Man's Extension into the Sea, MTS, January 1966, p. 111.

20. Rechnitzer, "Manned Deep Submersibles as Functional Systems", ASNE Journal, 1963, p. 455.

21. Williams, "Marine Completion Systems", Exploiting the Ocean, MTS, June 1966, p. 251.

22. Bartlett, "Marine Engineering and the Challenge by the Deep Frontier", Marine Technology, Vol. 2, No. 2, April 1965, p. 142.

23. Cruickshank, M., "Methods of Exploring the Ocean", Undersea Technology, Jan. 1964, p. 29.

24. Miller, Tirey and Mecarini, "Mechanics of Marine Mineral Exploration", ASME Paper No. 67-UNT-2, Underwater Tech. Conf., May 1967.

25. "Methods of Mineral Recovery", Undersea Technology, April 1964, p. 29.

26. Jackson, "Military Significance of Deep Diving Vehicles", ASNE Journal, 1963, p. 365.

27. Pehrson, "Mining Industry's Role in Development of Undersea Mining", Exploiting the Ocean, MTS, June 1966, p. 182.

28. "New Devices for Deep Sea Operations", Undersea Technology, April 1963, p. 24.

29. Kastrop, J. E., "New Drilling Technology Certain from Project Mohole", Petroleum Engineer, July 1964, p. 43.

30. "North Sea: Europe's Major Hope for New Petroleum Reserves", World Oil, August 15, 1964, p. 74.

31. Kim, H. C., "Note on Floodable and Subdivision of Oil Drilling Rigs", SNAME, Gulf Section, April 14, 1967 and Marine Technology, Vol. 5, No. 3, July 1968, p. 267.

SELECTED REFERENCES

Background Material

B. General Design Considerations

32. Corll, "Pressure Terminology", ASME Paper 65-WA/PT-2.

33. Welling and Cruickshank, "Review of Available Hardware Needed for Undersea Mining", Exploiting the Ocean, MTS, June 1966, p. 79.

34. Moore, D. G. and Palmer, H. D., "Offshore Seismic Reflection Surveys", Civil Eng. in the Oceans, ASCE, San Francisco, Sept. 6-8, 1967.

35. McNott, "Seismic Search for Offshore Oil", ASME Underwater Tech. Conf., May 1967.

36. Hales, "Single Ship Seismic Refraction Studies at Sea", ASME Underwater Tech. Conf., May 1967.

37. Webb, "Technology of Sea Diamond Mining", Ocean Sci. and Ocean Eng., Vol. I, MTS, 1965, p. 12.

38. McEnter, J. R., Geer, R. L. and Collipp, B. G., "Underwater Drilling and Completion Methods", AIME Soc. of Petroleum Engineers, Paper No. SPE755.

39. Picard, J., "Exploring the Gulf Stream", Oceanology International, May/June 1968, p. 28.

40. Busby, R.F., Hunt, L.M. and Rainie, W.O., "Hazards of the Deep", Ocean Industry, Vol. 3, No. 7, Part I, July 1968, p. 72.

41. Rudnick, P., "Motion of a Large Spar Buoy in Sea Waves", Journal of Ship Research, Dec. 1967, p. 257.

42. Vytlacil, N., "Pitch and Heave Response in Waves of a Spar-Shaped Hydrophone Platform", Lockheed Underwater Missile Facility Report TM81-7312, July 7, 1961.

43. Michel, "Sea Spectra Simplified", SNAME, Gulf Section, April 1967.

44. Ewing, M. and Engel, L., "Seismic Shooting at Sea", Scientific American, May 1962, p. 116.

SELECTED REFERENCES

Background Material

C. Structural Loadings

1. Chappelear, J. E., "Calculation of Wave Forces on a Sunken Obstruction," AIME, Petroleum Trans., Vol. 210, 1957, p. 227.

2. Wilson, B. W., "Case of Critical Surging of a Moored Ship", ASCE, Waterways and Harbors Div. Paper No. 2358, Dec. 1959.

3. Ippen, A. T., Estuary and Coastline Hydrodynamics, (p. 341), McGraw Hill Book Company.

4. Reid, R. O., "Correlation of Water Level Variations with Wave Forces on a Vertical Pile for Nonperiodic Waves", 6th. Intl. Conf. on Coastal Engineering, Dec. 1957, p. 749.

5. Hromadik, J. J. and Breckenridge, R. A., "Construction Concepts for the Deep Ocean", Civil Eng. in the Oceans, ASCE San Francisco, Sept. 6-8, 1967.

6. Reid, R. O., "Design Criteria for Fixed-Piling Structures and Mobile Units", API, DPP, 1954, p. 248.

7. Freeman, "Distribution of Winds, Tides, Waves and Standard Wave Forces in Hurricane Flossie of 1956", SNAME, Gulf, Feb. 15, 1957.

8. Laird, A.D.K. and Johnson, C. A., "Drag Forces on an Accelerated Cylinder, Journal of Petroleum Technology, May 1956, p. 65.

9. Verrett, "Effect of Shape and Depth on Wave Forced Oscillations of Submerged Moored Objects", M.I.T. S.M. Thesis (XIII-A) 1960, (Sept.)

10. Morison, J. R., O'Brien, M. P., Johnson, J. S. and Schaaf, S. A., "Force Exerted by Surface Waves on Piles", AIME, Petroleum, Vol. 189, 1950, p. 149.

11. O'Brien, J. T. and Kuchenreuter, D. I., "Forces Induced on a Large Vessel by Surge", ASCE, Waterways and Harbors Div. Paper No. 1571, March 1958.

12. Wilson, B. W., "Forces Induced on a Large Vessel by Surge", ASCE, Waterways and Harbors Div. Paper No. 1884-9, Dec. 1958.

13. Graham, J. R. and Pike, J. E., "Hurricane Flossie - Knowledge and Experience Gained for Offshore Drilling Equipment", API, Paper No. 926-2-B, Shreveport, March 1957.

14. Graham, J. R. and Pike, J. E., "Hurricane Flossie - Knowledge and Experience Gained for Offshore Drilling Equipment", SNAME, Gulf, Feb. 15, 1957.

SELECTED REFERENCES

Background Material

C. Structural Loadings

15. Gaul, R. D., "Model Study of a Dynamically Laterally Loaded Pile",
ASCE, Journal of the Soil Mechanics and Foundations Div. Paper No.
1535, Feb. 1958.

16. Wiegel, R. L., Oceanographical Engineering, p. 248, Prentice-Hall, Inc.

17. Borgman, L. E., "Ocean Wave Simulation for Engineering Design",
Civil Eng. in the Oceans, ASCE San Francisco, Sept. 6-8, 1967.

18. McClelland, B., Focht, J. A., and Enrich, W. J., "Problems in
Predeterming the Capacity of Long Piles", Civil Eng. in the Oceans,
ASCE San Francisco, Sept. 6-8, 1967.

19. Harrer, P. L., "Wave Forces Computed for a Typical Offshore Drilling Site", Amer. Inst. of Mining and Metallurgical Engineers,
San Antonio, Oct. 5-7, 1949.

20. Bretschneider, C. L., "Selection of Design Wave for Offshore
Structures", ASCE Waterways and Harbors Div., Paper 1568, March 1958.

21. Nolan and Honsinger, "Wave Induced Vibrations of Offshore Structures",
SNAME, Vol. 71, 1963, p. 680.

22. Freeman, J. C. and Gade, H., "Winds, Tides, Waves and Wave Forces
in Hurricane Flossie of 1956", API, Paper No. 926-2-C, Shreveport,
March 1957.

23. Morison, J. R. and O'Brien, M. P., "Forces Exerted by Waves on
Objects", Trans. of American Geophysical Union, Vol. 33, No. 1,
Feb. 1952.

24. Matlock, H. and Reese, L. C., "Foundation Analysis of Offshore
Pile-Supported Structures", Fifth International Conf., International
Society of Soil Mechanics and Foundation Eng., Paris, July 1961.

25. Michel, W. H., "How to Calculate Wave Forces and their Effects",
Ocean Industry, Vol. 2, No. 5, May 1967, pg. 58 and Vol. 2, No. 6,
June 1967, p. 49.

26. Peyton, H. R., "Ice and Marine Structures", Ocean Industry, Vol. 3,
No. 3, March 1968, p. 40, Vol. 3, No. 9, Sept. 1968, p. 59.

27. Pierson, W. J. and Holmes, P., "Irregular Wave Forces on a Pile",
Journal of the Waterways and Harbors Div., ASCE, Vol. 91, No. WW4,
November 1965.

28. LeMéhaute, B., "On Froude-Cauchy Similitude", Coastal Engineering
Santa Barbara Specialty Conference, October 1965.

SELECTED REFERENCES

Background Material

C. Structural Loadings

29. Bretschneider, C. L., "Probability Distribution of Wave Forces", Journal of the Waterways and Harbors Div., ASCE, Vol. 93, May 1967.

30. Dean, R. G., "Relative Validities of Classical and Numerical Water Wave Theories", Civil Eng. in the Oceans, ASCE, San Francisco, Sept. 6-8, 1967.

31. Borgman, L. E., "Spectral Analysis of Ocean Wave Forces on Piling", Journal of the Waterways and Harbors Div., ASCE, Vol. 93, No. WW2, May 1967, p. 129.

32. Dean, R. G., "Stream Function Representation of Non-Linear Ocean Waves", Journal of Geophysical Research, 70:18, Sept. 15, 1965.

33. Reid, R. O. and Bretschneider, C. L., "Surface Waves and Offshore Structures", A & M College of Texas, Oct. 1953.

34. Goda, Y., "Wave Forces on a Vertical Circular Cylinder", Port and Harbor Technical Research Inst., Ministry of Transportation, Japan, August 1964.

35. Jen, Y., "Wave Forces on Circular Cylindrical Piles used in Coastal Structures", Ph.D. Thesis, University of California, Berkeley, Cal., January, 1967.

36. MacCamy, R. C. and Fuchs, R. A., "Wave Forces on Piles: A Diffraction Theory", U. S. Beach Erosion Board, Tech. Memo. No. 69, Dec. 1954.

37. Borgman, L. E., "Wave Forces on Piling for Narrow-Band Spectra", Journal of the Waterways and Harbors Div., ASCE, Vol. 91, No. WW3, August 1965, Part I.

38. Brater, E. F., McNown, J. S. and Stair, L. D., "Wave Forces on Submerged Structures", Journal of the Hydraulic Div., ASCE, Vol. 84, HY6, November 1958, Paper No. 1833.

39. Wilson, B. W., "Design Sea and Wind Conditions for Offshore Structures", Proceedings of OECON, 1966, p. 665.

40. Skjelbreia, L., Hendrickson, J.A. and Kilmer, R.E., "Wave Force Calculation for Three-Dimensional Structures Composed of Tubular Members", Proceedings of OECON, 1966, p. 235.

SELECTED REFERENCES

Background Material

C. Structural Loadings

41. Geminder, R., "Ice Force Measurement", Proceedings of OECON, 1968, p. 665.

42. Beckman, H. and McBride, C. M., "Inherent Scatter of Wave Forces on Submerged Structures", ASME Paper No. 68-PET-7.

SELECTED REFERENCES

Buoyancy Elements and Systems

1. Hobaica, "Buoyancy Systems for Deep Submergence Structures", ASNE Journal, 1964, p. 733.

2. Stixrud, T. E., "Economical Deep Water Buoyancy", Undersea Technology, October 1964, p. 21, or ASNE Journal, August 1965, p. 593.

3. Gross, "Low-Cost Buoyant Element for Deep-Submergence Applications", Undersea Technology, March 1966, p. 23, or ASNE Journal, March 1966, p. 590.

4. Winer, "Syntactic Flotation Material", BuShips Journal, December 1965, p. 15.

5. Winer, "Syntactic Flotation Material for Deep Submergence Behicles", Ocean Sci. and Ocean Eng., Vol. I, MTS 1965, p. 623.

6. Resnick, I., and Macander, A., "Syntactic Foams for Deep Sea Engineering Applications", Civ. Eng. in the Oceans, ASCE San Francisco, Sept. 6-8, 1967 and Naval Engineer's Journal, Vol. 80, No. 2, April 1968, p. 235.

7. Irgon, "Use of Factor Dependence Analysis in the Selection of Buoyant Materials and Structures", Ocean Sci. and Ocean Eng., Vol. II, MTS 1965, p. 1212.

8. Waite, W. A., Waldron, M. L. and Nahabedian, A., "Thermoplastic Syntactic Foam for Structural Void Fillers", Undersea Technology, Vol. 9, No. 6, June 1968, p. 33.

9. Stechler, B. G. and Poneros, G. J., "Parametric Analysis of Optimum Buoyancy Module Designs with Computer Applications", Ocean Engineering, Vol. 1, No. 1, July 1968, p. 17.

SELECTED REFERENCES

Cable and Mooring Systems

1. Brainard, "Braiding Techniques Applied to Oceanographic Cables', MTS The New Thrust Seaward, p. 631.

2. Schneider and Nickels, "Cable Equilibrium Trajectory in a Three Dimensional Flow Field", ASME Paper 66-WA/UmT-12.

3. Jones, R. E., "Coast Guard Buoy Installation", Ocean Science and Ocean Eng., Vol. I, MTS, 1965, p. 259.

4. MacNaught, D. F., "Design of Cargo Handling Gear", SNAME, N. E. Section May 1947.

5. Schick, "Design of a Deep Moored Oceanographic Station", Buoy Technology (Supplement), MTS March 1964, p. 43.

6. Symonds & Trowbridge, "Development of Beam Trawling in the North Atlantic", SNAME, Vol. 55, 1947, p. 359.

7. "Dynamics and Kinematics of the Laying and Recovery of Submarine Cable", ASNE Journal, 1958, p. 531.

8. Poffenberger, Capadona, and Siter, "Dynamic Testing of Cables", Exploiting the Ocean, MTS, June 1966, p. 485.

9. "Improved Towline Design for Oceanography", Undersea Technology, May 1965, p. 57.

10. Brainard and Paney, "Low Noise Hydrodynamic Depth Stabilization of Towed Seismic Hydrophone Arrays", Proceedings of OECON 1967, p. 719.

11. McNeely, "Marine Fish Harvest Methods - Recent Advancements and Future Engineering Needs", ASME Underwater Tech. Conf. May 1967.

12. Alverson and Schaefers, "Methods of Search and Capture in Ocean Fisheries", Exploiting the Ocean, MTS, June 1966, p. 319.

13. Haas, "Natural and Synthetic Cordage in the Field of Oceanography", MTS, The New Thrust Seaward, p. 617.

14. "Nylon Rope for Deep-Sea Instrumentation", Undersea Technology, May 1963, p. 18.

15. Alverson and Schaefers, "Ocean Engineering - Its Application to the Harvest of Living Resources", Ocean Sci. and Ocean Eng., Vol. I, MTS, 1965, p. 158.

SELECTED REFERENCES

<u>Cable and Mooring Systems</u>

16. Jones, R. E., "Polypropylene and Nylon Rope", <u>Ocean Sci. and Ocean Eng.,</u> Vol. I, MTS 1965, p. 243.

17. Savastano, "Selecting Wire Rope for Oceanographic Applications", <u>Undersea Technology</u>, Feb. 1967, p. 21. (Nav. Eng. Journal Vol. 79, No. 3, June 1967, p. 480).

18. MacNaught, D. F., "Stays", <u>Design and Construction of Steel Merchant Ships</u>, Arnott (SNAME), p. 345.

19. "Stress Analysis of Ship-Suspended Heavily Loaded Cables for Deep Underwater Emplacements", Project "Trident" Technical Report by A. D. Little, Inc., for the Bureau of Ships.

20. "Synthetic Rope Use Gains in Buoy Systems", <u>Undersea Technology</u>, June 1965, p. 17.

21. McLoad and Bowers, "Torque Balanced Wire Rope and Armored Cables", <u>Buoy Technology</u>, MTS, March 1964, p. 233, or p. 341.

22. Schneider, Burton, and Maban, "Tow-Cable Snap Loads", <u>Marine Technology</u>, Vol. 2, No. 1, Jan. 1965, p. 42.

23. MacNaught, D. F., "Wire Ropes", <u>Design and Construction of Steel Merchant Ships</u>, Arnott (SNAME) p. 324.

24. Ogg, R. D., "Anchors and Anchoring", Danforth Anchors, 2121 Allston Way, Berkeley, California.

25. Smith, J. E., "Buoys and Anchorage Systems (Chap. 7 of "Structures in Deep Ocean Engineering", "Manual for Underwater Construction"), U. S. Naval Civil Eng. Lab. AD473928.

26. Beck, H. C. and Ess, J. D., "Deep Sea Anchoring", Hudson Laboratories, Columbia University Tech. Report 12-98.

27. Burchard, S., "Deep Sea Cables", <u>Oceanology International</u>, May/June 1968, p. 31.

28. <u>Design Manual - Harbor and Coastal Facilities</u>, U.S.N. Bureau of Yards and Docks, Navdocks DM-26,
           I   Harbors
          II   Coastal Protection
        III   Dredging
         IV   Ship Channels
          V   Fixed Moorings
         VI   Fleet Moorings
       VII   Mooring Design - Physical and Empirical Data

SELECTED REFERENCES

Cable and Mooring Systems

29. Morrow, B. W. and Chang, W. F., "Determination of the Optimum Scope of a Moored Buoy", Journal of Ocean Technology, Vol. 2, No. 1, Dec. 1967, p. 37.

30. Berteaux, H. O.; Capadona, E. A.; Mitchell, R. and Morey, R. L., "Experimental Evidence on the Modes and Probable Causes of a Deep-Sea Buoy Mooring Line Failure", Critical Look at Marine Technology, MTS, 1968, p. 671.

31. "Instructions for the Design and Care of Wire Rope Installations", U.S.N. BuShips Technical Bulletin No. 5.

32. Dove, H. L., "Investigations on Model Anchors", Trans. I.N.A., Vol. 92, 1950, p. 351.

33. Booda, L. L., "Major Buoy Programs", Undersea Technology, Sept. 1967, p. 24.

34. McNeely, R. L., "Marine Fish Harvest Methods - Recent Advancements and Future Engineering Needs", Journal of Ocean Technology, Vol. 2, No. 2, April, 1968, p. 2.

35. Petre, J. W., "Monster Buoys for Offshore Exploration", Suppl. to Proceedings of OECON, 1967, p. 76.

36. Wiegel, R. L., Oceanographical Engineering, Prentice-Hall, Inc., p. 487.

37. Frassetto, R., "Progress on a Compact Multipurpose and Economical Buoy System Having Low Hydro and Aerodynamic Drag", Journal of Ocean Technology, Vol. 2, No. 2, April 1968, p. 17.

38. Lindquist, C. E., "Rigging of Ships and Cargo Handling Gear", N. E. Section SNAME, March 28, 1944.

39. Chmelik, F. B., "Sediment Probes and Anchor System for Submersible Research Vehicles", Critical Look at Marine Technology, MTS, 1968, p. 503.

40. Smith, P. F., "Summary of Recent Deep Ocean Scientific Buoy Performance", NATO Subcommittee on Oceanographic Research Tech. Report #19.

41. "Technical Summary of Five Deep Sea Moors", Hydrospace Research Corp. Tech. Report # 11-1964.

42. Wire Rope Engineering Handbook, American Steel and Wire Co.

43. Money, "Anchors and Cable", SNAME, Philadelphia, Dec. 20, 1946.

44. Rennie, "Anchor Chain Cables", SNAME, Pacific N. W., Oct. 10, 1959.

SELECTED REFERENCES

Cable and Mooring Systems

45. Thompson, J. W., "Anchor, Mooring and Towing Arrangements", Design and Construction of Steel Merchant Ships, Arnott (SNAME), p. 348.

46. "ASR Deep Moor", NavShip Systems Command Technical News, Jan. 1967, p. 29.

47. Uyeda, "Buoy Configuration Resulting from Model Tests and Computer Study", Buoy Technology (Supplement), MTS, March 1964, p. 31.

48. Danforth, "Cables and their Effect on Anchor Loads", SNAME, N. California, April 13, 1945.

49. Wilson, B. W., "Characteristics of Deep-Sea Anchor Cables in Strong Ocean Currents", Texas A. & M. College, Dept. of Oceanography and Meterology, Feb. 1961.

50. Denton and Stahl, "Concept and Design of the SEAS Buoy System", MTS, The New Thrust Seaward, p. 493.

51. Jones, "Deep Ocean Installations", Ocean Sci. and Ocean Eng., Vol. I, MTS, 1965, p. 200.

52. Lewis and Lorenzen, "Deep Sea Mooring", Mechanical Engineering, Sept. 1966, p. 51, Vol. 88, No. 9.

53. Sherwood, W. G., "Deep Sea Mooring System", Undersea Technology, June 1967, p. 14.

54. "Deep Sea Moor - TOTO II", Nav. Ship Systems Command Technical News, Jan. 196 p. 35.

55. Shipley, H. P. and Brown, D. F., "Deep Sea Subsurface Mooring of a Model Submarine Hull", Civil Eng. in the Oceans, ASCE, San Francisco, Sept. 6-8, 1967.

56. "Deep Sea Terminals", Offshore, Sept. 1958, p. 47.

57. Lewis and Lorenzen, "Design of a Test System for Deep Ocean Research", ASME Paper 65-WA/UmT-9.

58. Metzler and Olsen, "Deep Ocean Surface Wave Measuring Buoy System", The New Thrust Seaward, MTS, p. 409.

59. Pritzlaff and Laniewski, "Development of a Self-Contained Deep Moored Buoy System", Buoy Technology, MTS, March 1964, p. 73 and ASNE Journal, Feb. 1965, p. 101.

60. Sterrett, E., "'Discoverer' puts Unique Mooring System to Work", Drilling, May 1963, p. 40.

SELECTED REFERENCES

Cable and Mooring Systems

61. Wilson, B. W. and Garbaccio, D. H., "Dynamics of Ship Anchor Lines in Waves and Currents", Civil Eng. in the Oceans, ASCE, San Francisco, Sept. 6-8, 1967.

62. Foster, "Five Year's Experience with Dynamic Anchoring", SNAME, Los Angeles Section, 1967.

63. Chapman and Rothenberg, "Hydrophone Suspension System for Deep Water Noise Measurement", Buoy Technology, MTS, March 1964, p. 199.

64. Farrell, "Improvements in Mooring Anchors", Trans. INA, Vol. 92, 1950, p. 335.

65. Smith, "Investigation of Embedment Anchors for Deep-Ocean Use", ASME Paper 66-PET-32.

66. Towne, "Mooring Anchors", Trans. SNAME, Vol. 67, 1959, p. 290.

67. Tudor, W. J., "Mooring and Anchoring of Deep Ocean Platforms", Civil Eng. in the Ocean, ASCE, San Francisco, Sept. 6-8, 1967.

68. Daubin, "Mooring Design and Performance of Oceanographic Buoys", Buoy Technology, MTS, March 1964, p. 58.

69. "Mooring for Artemus Installation Ship", Nav. Ship Systems Command Technical News, Jan. 1967, p. 31.

70. "Mooring Guide", U.S.N. Bureau of Yards and Docks, Publ. No. Nav. Docks TP-PW-2, 1 March 1954.

71. Graham, "Mooring Techniques in the Open Sea", Marine Technology, Vol. 2, No. 2, April 1965, p. 132.

72. Gay, "New Engineering Techniques for Application to Deep-Water Mooring", ASME Paper 66-PET-31.

73. Berteaux, H. O. and Fofonoff, N. P., "Oceanographic Buoys Gather Data from Surface to Sea Floor", Oceanology, July/Aug. 1967, p. 39.

74. Knapp, R. P. and Wait, H. V., "Offshore Mooring of Drilling Rig Tenders", ASME Paper No. 55-PET-26.

75. Thorpe and Farrell, "Permanent Moorings", Trans. INA, Vol. 90, 1948, p. 111.

76. Hobbs, M. X., Jr., "Shell Perfects New Submarine Completion Technique", World Oil, Sept. 1962, p. 101.

77. "Single-Buoy Mooring System Rates High with Shell", Oil and Gas Journal, Nov. 16, 1964, p. 194.

SELECTED REFERENCES

Cable and Mooring Systems

78. Daubin and Potter, "Some Design Criteria for Deep Sea Moorings",
     ASME Paper 63-WA-211.

79. "Standards of Design for Moorings", U.S.N. Bureau of Yards and Docks,
     Publ. No. NavDocks P-168, April 1945.

80. Cox, Johnson, Sandstrom and Jones, "Taut-Wire Mooring for Deep
     Temperature Recordings", Ocean Sci. and Ocean Eng., Vol. II, MTS, 1965.

81. Gerard, "Taut-Wire Navigation Buoys Used in the THRESHER Search",
     Buoy Technology (Supplement), MTS, March 1964, p. 57.

82. Falwell, Maddux and Triplett, "Tide Measuring Buoy System", Ocean Sci.
     and Ocean Eng., Vol. I, MTS, 1965, p. 313.

83. Gay, S. N., "Computer Analysis and Design of Undersea Cable Systems",
     Undersea Technology, Vol. 9, No. 10, Oct. 1968, p. 43.

84. Giorgi, E., "Underwater Power Systems", MTS, Journal of Ocean Technology,
     Vol. 2, No. 4, Oct. 1968, p. 57.

SELECTED REFERENCES

Diving Systems and Man in the Sea

1. Lindbergh, J. M., "Diver Assistance in Offshore Drilling", World Oil, June 1964, p. 133.

2. "Diving Equipment and Systems", NavShip Systems Command Technical News, Jan. 1967, p. 6.

3. Young, K. G., "Economic 600 ft. Water Operations are Expected", World Oil, April 1965, p. 151.

4. Emerson, "Future Applications of Prolonged Submergence Diving", Proceedings of OECON 1967, p. 553.

5. Emerson, O'Neill and Amskiewicz, "Life Support for Prolonged Submergence", ASME Paper No. 67-UNT-9, Underwater Tech. Conf., May 1967.

6. Schempf, J., "New Challenge - Divers Face Deeper Depths in Oil Search", Offshore, August 1967, p. 49.

7. Lindbergh, J. M., "New Developments in Deep Water Operations", API, Div. of Production, Los Angeles, April 29-30, 1964, Paper No. 801-40C.

8. Major, "Operational Tests and Installation of the Prototype Reading and Bates Diving System", Proceedings of OECON 1967, p. 453.

9. "Physiological Effects of Deep Diving", Offshore, August 1967, p. 58.

10. Campise, "Today's Diving Contractor: His Role in Engineering the Peacetime Ocean", ASME Paper No. 67-UNT-6, Underwater Tech. Conf. May 1967.

11. Folbert and Dowling, "Underwater Work Tools for Scientific Diving", ASME Paper No. 67-UNT-1, Underwater Tech. Conf. May 1967.

12. Craven, J. P., "Working in the Sea", Int'l. Sci. and Technology, April 1967, p. 50.

13. Munro, A. C., "Case for Diving Chambers", Offshore, Dec. 1967, p. 60.

14. Keller, H. and Buhlman, A. A., "Deep Diving and Short Decompression by Breathing Mixed Gases", Journal of Applied Physiology, Vol. 20, 1965, p. 1267.

15. Fonda-Bonardi, B. and Buckley, C. P., "Diving Suit from Space", Ocean Industry, Vol. 2, No. 9, Sept. 1967, p. 57.

16. Searle, W. F., "History of Man's Deep Submergence", U. S. Naval Institute Proceedings, Vol. 92, No. 3, March, 1966.

SELECTED REFERENCES

Diving Systems and Man in the Sea

17. Bowen, H. M., "Human Capability in Water", Critical Look at Marine Technology, MTS, 1968, p. 265.

18. Hoover, G. N., "Life Support Requirements", Ocean Industry, Dec. 1967, Vol. 2, No. 12, p. 24.

19. Wallman, H. and Kinsey, J. L., "Life-Support Systems for Undersea Use", Oceanology International, Jan/Feb. 1968, p. 31.

20. Streit. H. J., "Lightweight Diver Haven", Critical Look at Marine Engineering, MTS, 1968, p. 311.

21. Lambertsen, C. J., "Limitations and Breakthroughs in Manned Undersea Activity", Exploiting the Ocean (Supplement), MTS, 1966, p. 115.

22. Hill, E. C., "Performing Identical Tasks in Air and Underwater", Ocean Industry, Vol. 3, No. 7, July 1968, p. 62.

23. Bagnall, F., "Saturation and Conventional Diving Contrasted", Offshore, May 1968, p. 132, Vol. 28, No. 5.

24. Blanchard, F. A., "Some Engineering Aspects of a Scientific Undersea Laboratory", ASME Paper 67 WA/UNT-9.

25. Thomas, L. R., "Submersible Electric Motors", Oceanology International, Jan/Feb. 1968, p. 34.

26. Tolbert, W. H., Dowling, G. B., "Tools for the Scientific Diver", Mechanical Eng., May 1968, p. 23.

27. "Underwater Accidents and their Prevention", Journal of Ocean Technology, Vol. 2, No. 3, July 1968, p. 45.

28. Cayford, J. E., Underwater Work , Cornell Maritime Press, Inc.

29. Stang, P. R. and Wiener, E. L., "Working Diver" Performance in Cold Water", Critical Look at Marine Technology, MTS, 1968, p. 289.

30. Carpenter, E.P. and Cooke, T.S., "Logistic and Surface Support (Sealab III)", Undersea Technology, Vol. 9, No. 8, August 1968, p. 46.

31. Young, K.G. and Major, E.L., "Design and Development of a Submersible Work Chamber", Proceedings of OECON, 1966, p. 613.

SELECTED REFERENCES

Diving Systems and Man-in-the-Sea

32. Hunley, W. H., "Deep-Ocean Work Systems", Ocean Engineering, John Wiley and Sons, Inc. (Chapter 14) p. 493.

33. Black, M. D., "Divcon Submersible Diving Systems", Proceedings of OECON, 1968, p. 619.

34. Sachse, C. D., "Mark II Deep Diving System", MTS, Journal of Ocean Technology, Vol. 2, No. 4, Oct. 1968, p. 37.

35. Shenton, E. H., "Toward Deeper Water", Proceedings of OECON, 1968, p. 685.

SELECTED REFERENCES

Fabrication

1. Gayer, "Deepstar Hull Fabrications", _Marine Technology_, Vol. 2, No. 4, Oct. 1965, p. 360.

2. Moldenhauer, Stachiw, Tsuji and Stowell, "Design, Fabrication and Testing of Acrylic Pressure Hulls for Manned Vehicles", ASME Paper 65-WA/UNT-10.

3. Jue and Giannoccolo, "_Sealab II_ Dished Heads Explosively Formed at Hunter's Point Division, San Francisco Bay Naval Shipyard", _Marine Technology_, Vol. 3, No. 1, January 1966, p. 99.

4. Opsahl, R. and Terrana, D. B., "How PX-15 Hull was Constructed", _Ocean Industry_, Vol. 2, No. 10, Oct. 1967, p. 35.

5. "Ocean Systems and Union Carbide Develop New Technique for Welding Underwater", _Ocean Industry_, Vol. 2, No. 10, Oct. 1967, p. 62.

6. "Techniques Employed in DSRV Hull Construction", _Ocean Industry_, Vol. 2, No. 10, Oct. 1967, p. 39.

7. Silva, E. A., "Welding Processes in the Deep Ocean", _Naval Engineers Journal_, Vol. 80, No. 4, August 1968, p. 561.

8. Garland, C., "Design and Fabrication of Deep-Diving Submersible Pressure Hulls", Transactions, S.N.A.M.E., 1968.

9. Mattavi, J. L. and Seibert, A. G., "Feasibility Evaluation of Boron Filament-Wound Pressure Vessel", ASME Paper No. 68-PVP-21.

SELECTED REFERENCES

Manipulators and Handling Gear

1. Lynch, "Analysis of Underwater Manipulation", Proceedings of OECON, 1967, p. 529.

2. Glickman and Hehn, "Conceptual Design of a Manipulator Device for Deep Diving Vehicles", Ocean Sci. and Ocean Eng., Vol. II, MTS, 1965, p. 1046.

3. Mosher, "Dexterity and Agility Improvement", ASME Underwater Technology Meeting, New London, Conn., May 5-7, 1965.

4. Hunley and Houck, "Existing Underwater Manipulators", ASME Paper 65-UnT-8.

5. Karinen, "Land-Based Remote Handling Background of Underwater Handling Equipment", ASME Paper 65-UnT-7.

6. Hunley and Houck, "Underwater Manipulators", Mechanical Engineering, March 1966, p. 35, Vol. 88, No. 3.

7. Jones, "Manipulator Systems, A Means for Doing Underwater Work", Naval Eng. Journal, Vol. 80, No. 1, Feb. 1968, p. 107.

8. Methot, D. A., "Navy ASR's Feature Western Gear Heavy Handling Equipment", Undersea Technology, Vol. 9, No. 6, June 1968, p. 28.

9. Wischhoefer, W. and Jones, R., "Submersible Manipulator Developments", Undersea Technology, Vol. 9, No. 3, March 1968, p. 22.

10. Rechnitzer, A. B., "Undersea Manipulation", Proceedings of OECON, 1966, p. 645.

SELECTED REFERENCES

Materials

1. Perry, H., "Argument for Glass Submersibles", Undersea Technology,
    September 1964, p. 31, or ASNE Journal, 1965, p. 79.

2. Breckenridge, R. A., and Haynes, H., "Behavior of Structural Elements
    in the Deep Ocean", Civil Eng. in the Oceans, ASCE, San Francisco,
    Sept. 6-8, 1967.

3. Ailer and Reinhart, "Corrosion of Aluminum Alloys", ASNE Journal,
    1964, p. 443.

4. Jacobson, "Design and Material Considerations for Deep Submergence
    Pressure Hulls", Paper No. 3944, Douglas Missile and Space Systems
    Division, June 1966.

5. Koechele, "Designing to Prevent Fatigue Failures", ASME Paper 65-MD-15.

6. Fried and Graner, "Durability of Reinforced - Plastic Structural
    Materials in Marine Service", Marine Technology, Vol. 3, No. 3, July 1966,
    p. 321.

7. Vicars, "Engineered Material - Honeycomb", Mechanical Engineering Sept.
    1965, p. 34, Vol. 87, No. 9.

8. DeHart and Pickett, "Evaluation of High-Strength Materials for Use in
    Severe Environments", ASME Paper 66-PET-25.

9. Heller, Fioriti and Vasta, "Evaluation of HY-80 Steel as a Structural
    Material for Submarines", ASNE Journal, Feb. and April, 1965, p. 29 & 193.

10. Kies, "Filaments for Reinforcement and the Applicability of Filament
    Wound Laminates for Deep Submergence Vehicles", Marine Technology,
    Vol. 3, No. 1, January 1966, p. 52.

11. "Filament Wound Housing for Deep Sea Research Equipment", Undersea
    Technology, Vol. 6, No. 2, February 1965, p. 26.

12. Myers and Fink, "Filament Wound Structural Model Studies for Deep
    Submergence Vehicles", ASNE Journal, 1965, p. 275.

13. "Filament Wound Vessels for Deep Sea Applications", Undersea Technology,
    Sept./Oct. 1962, p. 16 or ASNE Journal, 1963, p. 188.

14. "Future Materials for Deep Submergence Application", Naval Ship Systems
    Command Technical News, January 1967, p. 45.

15. "Glass and Ceramics for Underwater Vehicle Structures", Undersea Technology,
    Vol. 5, No. 1, January 1964, p. 44.

SELECTED REFERENCES

Materials

16. Buhl, Hom, and Willner, "Glass-Reinforced Plastics for Submersible Pressure Hulls", ASNE Journal, Vol. 73, No. 4, Oct. 1963, p. 827.

17. Petker, I., "Factors in Strength Measurements of Glass Roving Strands", ASNE Journal, February 1966, p. 119.

18. Wright, "Hollow Glass Filament Reinforced Plastic for Deep Submergence Vehicles", Ocean Sci. and Ocean Eng., Vol. II, MTS, 1965, p. 824.

19. Gross and Heise, "Low Cycle Fatigue Behavior of Internally Pressurized Boxes", ASME Paper 66-WA/Unt-1.

20. Gross, "Low Cycle Fatigue of Materials for Submarine Construction", ASNE Journal, 1963, p. 783.

21. Vicars, "Marine Application of Titanium Alloys", SNAME Bulletin, Oct. 1952.

22. Thompson and Logan, "Materials for Deep Submergence", Oceanology International, March/April 1967, p. 25.

23. Bernstein, "Materials for Deep Submergence Capsules", Ocean Sci. and Ocean Eng., Vol. I, MTS, 1965, p. 583.

24. Fioriti, Vasta and Starr, "Materials for Hydrofoils", ASNE Journal, 1963, p. 609.

25. Sorkin, "Materials for Submarine Hard Sea Water Systems", ASNE Journal, 1965, p. 93.

26. Frisby, D. L., "New Steels Solve Problems Offshore", Oil and Gas Journal, February 1, 1965, p. 62.

27. Hunt and Bellware, "Ocean Engineering Hardware Requires Copper - Nickel Alloys", MTS, The New Thrust Seaward, p. 243.

28. "Plastic Pressure Hulls", Undersea Technology, Vol. 2, No. 4, July/August 1961, p. 18.

29. Fried, Kaminatsky and Silvergleit, "Potential of Filament Wound, Glass Reinforced Plastics for Construction of Deep Submergence Pressure Hulls", ASME Paper 66-WA/UnT-9.

30. Levenetz, "Potential of Nonmetallics for Use as Structural Materials for Deep Submergence Vehicles", Ocean Sci. and Ocean Eng., Vol. I, MTS, 1965, p. 601.

31. Kim, Y. C., "Protective System of Concrete Structure Subjected to Underwater Shock Waves", Civil Engineering in the Oceans, ASCE, San Francisco, Sept. 6-8, 1967.

SELECTED REFERENCES

Materials

32. "Reinforced Plastics for Hydrospace Vehicles", ASNE Journal, 1961, p. 775.

33. Sandwich Cylinder Construction for Underwater Pressure Vessels", ASNE Journal, 1964, p. 951.

34. "Sea Water Corrosion of Engineering Materials", Undersea Technology, April 1963, p. 20.

35. Gerwick, B. C., Jr., "Techniques for Concrete Construction on the Ocean Floor", Civil Engineering in the Oceans, ASCE, San Francisco, Sept. 6-8, 1967.

36. "Titanium, A Survey", ASNE Journal, 1953, p. 549.

37. Green, Pipher and Ochieano, "Titanium Fabricating Techniques", ASME Paper 65-AV-20.

38. Minkler, W. M. and Feige, N., "Titanium for Deep Submergence Vehicles", Undersea Technology, January 1965, p. 26, or ASNE Journal, Vol. 77, No. 3, June 1965, p. 386.

39. Starr and Warkentin, "Titanium for High Speed Hydrofoils", SNAME, Hydrofoil Symposium, Seattle, 1965.

40. McKee, Lewis and Ballass, "Use of High Yield Steel in Submarine Construction", SNAME, N. Y. Metropolitan Section, April 23, 1957.

41. "Welding of High Strength Aluminum Shells", Undersea Technology, Vol. 4, No. 1, January 1963, p. 38.

42. "Hollow Core Structures for High Pressure Hulls", Undersea Technology, Vol. 2, No. 3, May/June 1961, p. 34.

43. Leveau, C. W., "Aluminum and its Use in Naval Craft", Naval Engineers Journal, Vol. 77, No. 1, Feb. 1965, p. 13.

44. Ahearn, R. L., "Consideration of Fracture Toughness in Pressure-Vessel Design", ASME Paper No. 67-MET-18.

45. Uhlig, H. H., Corrosion Handbook, John Wiley & Sons, New York.

46. Gross, M. R. and Czyryca, E. J., "Effects of Notches and Saltwater Corrosion on the Flexural Fatigue Properties of Steels for Hydrospace Vehicles", Naval Engineers Journal, Vol. 79, No. 6, Dec. 1967, p. 1003.

47. Richards, C. W., Engineering Materials Science, Wadsworth Publishing Co., San Francisco, 1961.

48. Forrest, P. G., Fatigue of Metals, Addison-Wesley Publ. Co.

SELECTED REFERENCES

Materials

49. Silvergleit, M., Kaminetsky, J. and Fried, N., "Filament Wound Glass Reinforced Plastics for Deep Sea Vehicles: The Present State-of-the-Art", SAMPE, Vol. 12, 1967.

50. Tuthill, A. H. and Schillmoller, C. M., "Guidelines for Selection of Marine Materials", Journal of Ocean Technology, Vol. 2, No. 1, Dec. 1967, p. 6.

51. Shankman, A. D., "Materials for Pressure Hulls - Present and Future", Ocean Industry, Vol. 3, No. 5, May 1968, p. 29.

52. Groves, "Materials in the Sea", Naval Engineers Journal, Vol. 80, No. 1, Feb. 1968, p. 35.

53. Wheatfall, W. L., "Metal Corrosion in Deep-Ocean Environments", Naval Engineers Journal, Vol. 79, No. 4, August 1967, p. 611.

54. Groves, Don, "Ocean Materials", Naval Engineers Journal, Vol. 80, No. 2, April 1968, p. 185.

55. Alfers, J. B. and Graner, W. R., "Reinforced Plastics - A Structural Material for Marine Applications", Trans. SNAME, Vol. 62, 1954, p. 5.

56. Pickett, A. G. and Grigory, S. C., "Prediction of the Low Cycle Fatigue Life of Pressure Vessels", ASME Paper No. 67-MET-3.

57. Elliott, D. R., MacDonald, D. C., Francois, E. Jr., Uhlig, E. C., "Radial Filament Spheres for Deep Submergence", SAMPE, Vol. 12, 1967.

58. Petrisko, "Stresses from Strain on Woven-Roving, Fiberglass Reinforced Plastic", Naval Engineers Journal, Vol. 80, No. 1, Feb. 1968, p. 95.

59. Forman, W. and DeHart, R., "Submersible 'Deep View' Pioneers Glass/Metal Bonding", Undersea Technology, Dec. 1967, p. 24.

60. Heller, S. R., "Use of Quenched and Tempered Steels for Welded Pressure Vessels", Naval Engineers Journal, Vol. 79, No. 5, Oct. 1967, p. 709.

61. Havens, F. E. and Bruner, J. P., "Fatigue of High Strength Structural Steels for Offshore Drilling Platforms", Proceedings of OECON, 1966, p. 373.

62. Cox, D. W., "Titanium goes Undersea", Undersea Technology, Vol. 9, No. 2, Feb. 1968, p. 31 and Naval Engineers Journal, Vol. 80, No. 4, Aug. 1968, p. 583.

63. Breidenbach, L. J., "Design of Plastic Structures for Deep Sea Use", ASME Paper No. 68-DE-55.

64. Comstock, J. M., "Uses of Titanium in Deep Submergence Vehicles", ASME Paper No. 68-DE-2.

65. Lebovits, A., "Permeability and Swelling of Elastomers and Plastics at High Hydrostatic Pressures", Ocean Engineering, Vol. 1, No. 1, July 1968, p. 91.

SELECTED REFERENCES

Materials

66. Woodland, B. T., "Structures for Deep Submergence", Space/Aeronautics,
     March 1967, p. 100.

67. Eager, W. J., "Sealab III Complex", Undersea Technology, Vol. 9, No. 8,
     Aug. 1968, p. 28.

68. LaQue, F. L., "Materials Selection for Ocean Engineering", Ocean Engineering,
     (Brahtz) John Wiley and Sons, Inc. (Chapter 16) p. 588.

69. Smith, C. A., "Velcoity Effects on the Corrosion Rates of Metals in
     Natural Seawater", Southeast Section S.N.A.M.E., May 10, 1968.

The following are reports of the U. S. Navy David Taylor Model Basin (now
Naval Ship Research and Development Center), Carderock, Md.

Feasibility Studies

70. Willner, A. R. and Sullivan, V. E., "Progress Report. Metallurgical
     Investigation of Titanium Alloys for Application to Deep-Diving
     Submarines", Report No. 1482, Dec. 1960.

71. Blumenberg, W. F., Hom, K., and Pulos, J. G., "Investigation of the Strength-
     Weight Characteristics of Cylindrical Sandwich-Type Pressure Hull Structures",
     Report No. 1678, May 1965.

72. Hom, K., Couch, W. P. and Willner, A. R., "Elastic Material Constants of
     Filament-Wound Cylinders Fabricated from E-HTS/E787 and S-HTS/E787 Prepreg
     Rovings", Report 1823, Feb. 1966.

73. Hom, K, and Couch, W. P., "Investigation of Filament Reinforced Plastics
     for Deep-Submergence Application", Report No. 1824, Nov. 1966.

74. Krenzke, M., Hom, K. and Proffitt, J., "Potential Hull Structures for
     Rescue and Search Vehicles of the Deep-Submergence Systems Project",
     Report No. 1985, March 1965.

75. Proffitt, J. L., "Hydrostatic Pressure Tests of Cylinders Fabricated from
     Hollow-Filament, Glass-Reinforced Plastic", Report No. 2132, Dec. 1965.

76. Pulos, J. G. and Krenzke, M. A., "Recent Developments in Pressure Hull
     Structures and Materials for Hydrospace Vehicles", Report No. 2137,
     Dec. 1965.

77. Kiernan, T. J., "An Exploratory Study of the Feasibility of Glass and
     Ceramic Pressure Vessels for Naval Applications", Report No. 2243,
     Sept. 1966.

SELECTED REFERENCES

Materials

Experimental Studies

78. Pulos, J. G. and Buhl, J. E. Jr., "Hydrostatic Pressure Tests of an Unstiffened Cylindrical Shell of a Glass-Fiber Reinforced Epoxy Resin", Report No. 1413, April 1960.

79. Krenzke, M. A., "Exploratory Tests of Long Glass Cylinders under External Hydrostatic Pressure", Report No. 1641, Aug. 1962.

80. Krenzke, M. A. and Kiernan, T. J., "Tests of Ring-Stiffened Composite Cylinders under External Hydrostatic Pressure", Report No. 1725, Sept. 1963.

81. Hom, K., Buhl, J. E. Jr., and Couch, W. P., "Hydrostatic Pressure Tests of Unstiffened and Ring-Stiffened Cylindrical Shells Fabricated of Glass-Filament Reinforced Fabrics", Report No. 1745, Sept. 1963.

82. Hom, K. and Blumenberg, W. F., "Hydrostatic Tests of Structural Models for Preliminary Design of a Web-Stiffened Sandwich Pressure Hull", Report No. 1763, Sept. 1963.

83. Couch, W. P. and Hom, K., "Cyclic Pressure-Loading Tests of a Ring-Stiffened Cylinder Fabricated of Glass-Filament Reinforced Plastic", Report No. 1825, May 1964.

84. Proffitt, J. L., "Hydrostatic Pressure Tests of Glass-Reinforced Plastic Cylinders with a Titanium Jacket", Report No. 1851, Jan. 1965.

85. Couch, W. P., "Hydrostatic, Creep, and Cyclic Tests of Radially Oriented Glass-Fiber Reinforced Plastic Spheres", Report No. 2089, Sept. 1965.

86. Nishida, K., "Static and Cyclic Fatigue Tests of Fusion-Sealed Pyrex Spheres", Report No. 2246, Sept. 1966.

87. Ward, G. D. and Hom, K., "Hydrostatic Pressure Tests of Glass-Reinforced Plastic Sandwich Cylinders with Lightweight Foam Core", Report No. 2291, Dec. 1966.

SELECTED REFERENCES

Offshore Drilling and Production Platforms

1. Little, "American Bureau of Shipping and Offshore Drilling", Proceedings of OECON, 1967, p. 54.

2. McDonald, R. W., "Analysis of Offshore Structure Design", Offshore, Feb. 1958, p. 35.

3. Reese, L. C. and O'Neil, M. W., "Analysis of Three-Dimensional, Pile-Supported Space Frames Subjected to Inclined and Eccentric Loads", Civil Eng. in the Oceans, ASCE, San Francisco, Sept. 6-8, 1967.

4. Bethlehem Mobile Platform, The Bethlehem Steel Co., Beaumont, Texas Booklet 383.

5. Hamlin, "Catamaran as a Seagoing Work Platform", Ocean Sci. and Ocean Eng., Vol. II, MTS, 1965, p. 1127.

6. Reese, L. C. and Johnston, L. P., "Criteria for the Design of Offshore Structures", Soc. of Petroleum Engineers of AIME, Paper No. SPE 483.

7. Eissler, V. C., Domingos, E. K. and Kearney, B. E., "Deep-Water Drilling in the Gulf of Mexico from a Floating Vessel", API Paper No. 926-8-E, March 1963.

8. Armstrong, "Description of Sun-Seadrome, Floating Offshore Drilling Rig", SNAME, Philadelphia, Feb. 27, 1951.

9. West, "Design and Construction of Offshore Oil Drilling Outfits", Institute of Marine Engineers, (Gt. Britain).

10. Geyer, R. A., "Design Criteria for Fixed-Piling Structures and Mobile Units", A.P.I., D.P.P., 1954, p. 244, (Offshore Operating Symp., Houston, March 1954.).

11. McClelland, B., "Design Criteria for Fixed Piling Structures and Mobile Units with Reference to Soil Mechanics", API, DPP, 1954, p. 249.

12. Collipp, B. G., "Design of an Offshore Mobile Drilling Unit", Shell Oil Co., Exploration and Production Dept. TS Report 65.

13. Strohbeck, E. E., "Design of Deep Water Structures for the Gulf of Mexico", Civil Eng. in the Oceans, ASCE San Francisco, Sept. 6-8, 1967.

14. Howe, R. J., "Design of Offshore Drilling Structures", ASME Paper 54-PET-19.

15. Howe, R. J., "Design of Offshore Drilling Structures", The Petroleum Engineer, Oct. 1955, p. B-73; Nov. 1955, p. B-77.

SELECTED REFERENCES

Offshore Drilling and Production Platforms

16. Rutledge, P. C., "Design of Texas Towers Offshore Radar Stations",
    Eighth Texas Conference on Soil Mechanics and Foundation Engineering,
    Sept. 14-15, 1956.

17. Howe, R. J., "Development of Offshore Drilling and Production Technology",
    ASME Underwater Tech. Conf., May 1967.

18. McClure, "Development of the Project Mohole Drilling Platform", SNAME.
    Vol. 73, 1965, p. 50.

19. Savage, G. M., "Discoverer", API, Paper No. 926-8-Jc, March 1963.

20. "Discoverer II", Offshore, June 20, 1967, p. 88.

21. Montgomery, "Drilling in the Sea from Floating Platforms", Exploiting
    the Ocean, MTS, June 1966, p. 230.

22. Hodgkinson, "Drilling Rig Construction", N.E.C.I.E.S., 6 March 1967.

23. "Drilling Rigs", American Bureau of Shipping, Publ. Dec. 15, 1964.

24. Shubinski, R. P., Wilson, E. L. and Selna, L. G., "Dynamic Response of
    Deep Water Structures", Civil Eng. in the Oceans, ASCE, San Francisco,
    Sept. 6-8, 1967.

25. Nath, J. H. and Harleman, D. R. F., "Dynamic Response of Fixed Offshore
    Structures to Periodic and Random Waves", Civil Eng. in the Oceans,
    ASCE, San Francisco, Sept. 6-8, 1967.

26. Mosby, R. C. and Greve, W. F., "Early Development of the FLoating
    Drilling Vessel for the Gulf of Mexico", API Paper No. 926-8-Ja,
    March, 1963.

27. Wiggins, "Earthquake Risk in Fixed Structure Design and Analysis",
    Proceedings of OECON 1967, p. 313.

28. Newmark, N. M., "Effect of Dynamic Loads on Offshore Structures",
    Eighth Texas Conference on Soil Mechanics and Foundation Engineering,
    Sept. 14-15, 1956.

29. Kampschaefer, Bruner and Havens, "Engineering Data for the Design and
    Fabrication of Offshore Drilling Platforms with Heat Treated Steels",
    ASME #65-PET-25, Sept. 19-22, 1965.

30. Rechtin, Steele and Scales, "Engineering Problems Related to the Design
    of Offshore Mobile Platforms", SNAME, Vol. 65, 1957, p. 633.

31. Howe, R. J., "Evaluation of Offshore Mobile Drilling Units",
    Ocean Industry, April 1966, p. 11.

SELECTED REFERENCES

<u>Offshore Drilling and Production Platforms</u>

32. Reese, L. C., "Factors Affecting the Stability of Offshore Drilling Platforms", API, Div. of Production Paper No. 926-7-H.

33. Anderson, Bartholomew & Wong, "Fixed Offshore Platform Design Analysis", Proceedings of OECON 1967, p. 277.

34. Temple, G., "Floating Drilling Vessel Design", ASME Paper No. 63-PET-16.

35. Seymour and McCardell, "Floating-Type Offshore Drilling Vessels", SNAME, Pacific Northwest 1967, also <u>Maritime Reporter</u>, April 1, 1967, p. 24 and <u>Marine Technology</u>, Oct. 1967, p. 388.

36. Sybert, J. H., "Foundation Scour and Remedial Measures for Offshore Platforms", Soc. of Petroleum Engineers of AIME, Paper No. SPE485.

37. "Giant Platform designed for Cook Inlet", <u>Oil and Gas Journal</u>, March 9, 1964, p. 60.

38. Rogers, L. C., "Giant Platforms Tame Cook Inlet", <u>Oil and Gas Journal</u>, Aug. 29, 1966, p. 70.

39. Atwood, J. H. and Stratton, H., "Global Marine's Floating Drilling Vessel", API Paper No. 926-8-Jb.

40. Rogers, L. C., "'Hilda' Kicks Operating Costs Up", <u>Oil and Gas Journal</u>, June 21, 1965, p. 154.

41. "History of Offshore Drilling", <u>Drilling</u>, March 1963, p. 44.

42. Bauer, R. F., Field, A. J. and Stratton, H., "How 'Cuss I' Drilled Below 6,000 ft.", <u>Offshore</u>, Dec. 1958, p. 20.

43. Denzler, H. E., "How Safety is Designed into Offshore Platforms", <u>World Oil</u>, June 1961, p. 131.

44. Grahn, R. H., "How to Select a Floating Rig", <u>Oil and Gas Journal</u>, Aug. 5, 1963, p. 99.

45. Kartinen, E. O., "Huntington Beach Offshore Platform 'Emmy'", API Div. of Production, Los Angeles April 29-30, 1964, Paper No. 801-40C.

46. Numata, "Hydrodynamic Model Tests of Offshore Drilling Structures", SNAME, <u>Marine Technology</u>, Vol. 3, No. 3, July 1966, p. 288.

47. Rogers, L. C., "Innovations Stir Interest in New Offshore Rig", <u>Oil and Gas Journal</u>, Aug. 30, 1965, p. 82.

48. "Kerr – McGee Orders First Jack-Up Rig", <u>Offshore</u>, Feb. 1965, p. 23.

SELECTED REFERENCES

Offshore Drilling and Production Platforms

49. Seale, "Marine Equipment for Offshore Oil crilling", SNAME, Gulf, Jan. 28, 1949.

50. "Marlin to Build Jack-Up Platform", Offshore, Sept. 1964, p. 19.

51. Wilson, W. G., "Meeting the Deepwater Challenge", Offshore, Feb. 1959, p. 24.

52. Bauer, R. F., Field, A. J. and Stratton, H., "Method of Drilling from a Floating Vessel and the 'Cuss I'", API, DPP, 1958, p. 124.

53. LeTourneau, R. L., "Mobile and Fixed Platforms for Waters Up to 600 Feet", ASME Paper No. 57-PET-11.

54. Macy, R. H., "Mobile Drilling Platforms", AIME Soc. of Petroleum Eng. Paper No. SPE1410.

55. "Mobile Drilling Units (Classed by Water-Depth Capacity)", Oil and Gas Journal, June 20, 1966, p. 127.

56. "Mobile Underwater Oil Storage Unit", Offshore, October 1960, p. 21.

57. Payne, J. M., "Mobile Units for Offshore Drilling", API, DPP, 1954, p. 257.

58. Gaucher, P. C., "Moving Drill Barges Safely", Offshore, April 1967, p. 94 and Ocean Industry, April 1967, p. 49.

59. Hurt & Sandberg, "New Concepts for Fixed Offshore Platforms", Proceedings of OECON 1967, p. 331.

60. Sumner, M. N., "New Ideas for Drilling Offshore", World Oil, July 1966, p. 135.

61. Rogers, L. C., "New Offshore Tool: Mobile Workover Rig", Oil and Gas Journal, Sept. 14, 1964, p. 90.

62. Mashburn, M. K. and Hubbard, J. L., "An Ocean Structure", Civil Eng. in the Oceans, ASCE, San Francisco, Sept. 6-8, 1967.

63. Pratt and Roy, "Offshore Drilling Barges", NECIES, Vol. 76, 1959-60, p. 157.

64. Kolodzey, "Offshore Drilling Platforms - Structural Details, Wave Forces and Safety Precautions", SNAME, Gulf, April 23, 1954.

65. "Offshore Drilling Practices", World Oil, May 1965, p. 110.

66. Okabe, S., "Offshore Drilling. Producing and Storage", Offshore, July 1958, p. 43.

SELECTED REFERENCES

Offshore Drilling and Production Platforms

67. "Offshore Drilling Rigs and Related Problems"
                Part 1 - Drilling from A Floating Vessel - Bauer, Crooks et al
                Part 2 - Fixed Problems - Shumate
                Part 3 - Artificial Islands - Pollard
                Part 4 - Towboat & Supply Vessels - Slocumbe
                Part 5 - Mooring - Rand
        SNAME, N. California, Oct. 11, 1958.

68. "Offshore Drillrigs", Lloyd's Register 100 A1 #17, 1966, p. 19.

69. Howe, R. J. and Collipp, B. G., "Offshore Mobile Units - Present and
     Future", Mechanical Engineering, Vol. 79, No. 4, April 1957, p. 335.

70. "Offshore Operations (Louisiana, Texas, California)", World Petroleum,
     November 1954.

71. "Offshore Production Practices", World Oil, May 1965, p. 124.

72. Cox, H. D., "Operation of Mobile Drilling Units Offshore", API,
     DPP, 1955, p. 68.

73. Friede, "Outline of Development of Offshore Drilling Equipment",
     ASNE Journal, 1955, p. 883.

74. "Plumbing the Seas for Oil", Fortune, Feb. 1965, p. 131.

75. Marshall, P. W., "Risk Factors for Offshore Structures", Civil Eng.
     in the Oceans, ASCE, San Francisco, Sept. 6-8, 1967.

76. Shields, C. M., "Role of Floating Platforms for Marine Drilling and Well
     Completion Operations", AIME Soc. of Petroleum Engineers, Paper No.
     SPE754.

77. Shields, C. M., "Role of Floating Platforms for Marine Drilling and
     Well Completion Operations", Sixth World Petroleum Congress, June 1963.

78. Grahn, R. H., "Selection of Floating Drilling Equipment", API Div. of
     Prod., Paper No. 801-39L, Los Angeles, May 21-23, 1963.

79. Kastrop, J. E., "Semi-Submersibles" New Breed of Floating Rigs",
     Petroleum Engineer, Oct. 1965, p. 94; Petroleum Engineer, Nov. 1965, p. 98.

80. Monney, "Slope Stability Factors to Consider in Offshore Drilling
     Operations", ASME Paper #67-UNT-3, Underwater Technology Conf., May 1967.

81. McClelland and Focht, "Soil Mechanics as Applied to Mobile Drilling
     Structures", SNAME, Gulf, Feb. 3, 1956.

82. Howe, R. J., "Some Factors in the Engineering Design of Offshore Mobile
     Drilling Units", API, DPP, 1955, p. 209.

SELECTED REFERENCES

Offshore Drilling and Production Platforms

83. Foster, E. T., "Statistical Prediction of Wave-Induced Responses of Deep Ocean Tower Structures", Civil Eng. in the Oceans, ASCE, San Francisco, Sept. 6-8, 1967.

84. "Stormdrill I Ready", Drilling, Aug. 1964, p. 70.

85. "Stormdrill V Ready for Work off Texas", Offshore, April 1967, p. 26.

86. Yachnis, M., "Structural Design Criteria for Fixed Deep Ocean Structures", Civil Eng. in the Oceans, ASCE San Francisco, Sept. 6-8, 1967.

87. Shields, C. M. and Shumate, N. B., "Summerland Offshore Drilling Platforms and their Differences", API, Div. of Production, Los Angeles, May 11-12, 1961, Paper No. 801-37G.

88. "This Tripod Rig Packs a Triple Punch", Offshore, Aug. 1964, p. 25.

89. Gibson, R., "Those Big Offshore Rigs" What They are and Why", World Petroleum, May 1965, p. 22.

90. "Three Years may be Required to Drill through Earth's Crust", Offshore, Oct. 1963, p. 23.

91. Chang, "Transverse Forces on Cylinders due to Vortex Shedding in Waves", S.M. Thesis (MIT) (Course XIII), Feb. 1964.

92. "Two More 'Floating Islands' for Odeco", Offshore, Sept. 1964, p. 27; March 1965, p. 45.

93. DeHart, R. C., "Underwater Vehicles and Deep Water Offshore Drill Platforms", Civil Eng. in the Oceans, ASCE, San Francisco, Sept. 6-8, 1967.

94. "Unique Mobile Storage Unit for Tenneco", Offshore, June 1965, p. 19.

95. Mosby, R. C., "What Now for Deep Water?", Drilling, March 1963, p. 43.

96. Shields, C. M., "What Stanocal has Learned about Drilling from Floating Platforms", Oil and Gas Journal, Aug. 1963, p. 48.

97. Laborde, A. J., "Why the Ocean Driller?", API, Paper No. 926-8-Ja, March 1963.

98. "Zapata's Maverick: New Concept in Deepwater Mobile Rigs", Oil and Gas Journal, Sept. 28, 1964, p. 71.

99. "Deepwater Package May Replace Fixed Platforms", Offshore, Vol. 28, No. 9, Aug. 1968, p. 64.

100. "Directory of Marine Drilling Rigs", Ocean Industry, Vol. 3, No. 3, March 1968, p. 19.

SELECTED REFERENCES

Offshore Drilling and Production Platforms

101. Macy, R. H., "Drilling Rigs", Ship Design and Construction (d'Arcangelo), SNAME, Chapter XVI.

102. Cooper, G. W., "Hurricane Damage to Structures in the Gulf of Mexico", Ocean Industry, Vol. 2, No. 10, Oct. 1967, p. 30.

103. "Latest Rig Concepts: Accent is on Mobility", Offshore, March 1968, p. 29.

104. "New Concepts for Deepwater Oil Production", Ocean Industry, Part I, Vol. 3, No. 8, Aug. 1968, p. 40, Part II, Vol. 3, No. 9, Sept. 1968, p. 48.

105. "New Drilling Rigs (for Neptune, Walker-Huthnance and Penrod)", Ocean Industry, Vol. 3, No. 5, May 1968, p. 5, 8, and 11.

106. Howe, R. J., "Offshore Mobile Drilling Units", Ocean Industry, Vol. 3, No. 7, July 1968, p. 38.

107. Lee, G. C., "Offshore Structures - Past, Present, Future and Design Considerations", Offshore, Vol. 28, No. 6, June 5, 1968, p. 45.

108. "Offshore with the Bureau", American Bureau of Shipping Annual Report 1967, p. 9.

109. Robertson, J. F., "Penrod Building Two Fast, Powerful Jack-Ups", Offshore, May 1968, Vol. 28, No. 5, p. 37.

110. "Probable Causes of 'Sea Gem' Disaster", Ocean Industry, Dec. 1967, Vol. 2, No. 12, p. 10.

111. "Sea Kills Two Rigs (Ocean Prince and Julie Ann)", Ocean Industry, Vol. 3, No. 4, April 1968, p. 10.

112. Willey, M. B., "Structures in the Sea", Petroleum Engineer, Nov. 1953, p. B-38.

113. Macy, R. H., "Trends in Offshore Drilling Rigs", Louisiana Engineering Society Annual Meeting, January 11, 1963.

114. Schempf, F. J., Jr., "Winter Storms Batter, Sink Two Offshore Rigs", Offshore, April 1968, p. 41.

115. Rosie, R. D., "Coast Guard's Offshore Lightstation Program", Naval Engineers Journal, Vol. 80, No. 4, August 1968, p. 609.

116. Reese, L. C. and Matlock, H., "Behavior of a Two-Dimensional Pile Group under Inclined and Eccentric Loading", Proceedings of OECON, 1966, p. 123.

117. Hauber, F. R., "Drilling and Production Structures for Oil and Gas on the Continental Shelf", Proceedings of OECON, 1966, p. 141.

118. St. Denis, M., "Analysis of the Dynamics of a Drilling Platform of the Jack-Up Type in the Floating Condition", Proceedings of OECON, 1966, p. 261

119. Mitchell, D. H., "Dynamic Environmental Loading Considerations for a Freely Floating Deep Water Drilling Rig", Proceedings of OECON, 1966, p. 395.

SELECTED REFERENCES

Offshore Drilling and Production Platforms

120. Moore, W. W. and Smoots, V. A., "Foundation Studies for Offshore Structures", _Proceedings of OECON_, 1966, p. 761.

121. Fowler, J. W., "Construction of the Chesapeake Light Station", _Civil Engineering_, Vol. 35, No. 1965, p. 76.

122. Daigle, J. B., "Cook Inlet Drilling and Production Platforms", _Proceedings of OECON_, 1968, p. 19.

123. Hromadik, J. J., "Deep Ocean Installations and Fixed Structures", _Ocean Engineering_ (Brahtz), John Wiley and Sons, Inc. (Chapter 10), p. 310.

124. Watts, J. S. and Faulkner, R. E., "Designing an Drilling Rig for Severe Seas", _Ocean Industry_, Vol. 3, No. 11, Nov. 1968, p. 28.

125. LeTourneau, R. L., "New Concepts in Worldwide Mobilization of Mobile Drilling Units", _Proceedings of OECON_, 1968, p. 49.

126. Lee, G. C., "Offshore Structures, Past, Present, Future and Design Considerations", _Proceedings of OECON_, 1968, p. 169.

127. "Rules for Building and Classing Offshore Mobile Drilling Units", _American Bureau of Shipping_, 1968.

128. Ruffin, J. V., "Steel Offshore Towers Replace Lightships", _Civil Engineering_, Vol. 35, Nov. 1965, p. 72.

SELECTED REFERENCES

Pipe Stresses

1. Bennett, "Capabilities and Applications of the Becker Drill for Offshore Sampling and Mining Explorations", ASME Underwater Tech. Conf., May 1967.

2. Graham, R. D., Frost M. A. and Wilhoit, J. C., "Analysis of the Motion of Deep-Water Drill Strings", (2 Parts), ASME Paper No. 64-PET-6 and ASME Paper No. 64-PET-7.

3. Matlock, H., "Applications of Numerical Methods to Some Structural Problems in Offshore Operations", Journal of Petroleum Technology, Sept. 1963, p. 1040.

4. "Deep-Sea Line is 'rehearsed'", Oil and Gas International, Vol. 1, No. 11, Nov. 1961, p. 42.

5. Wilkinson, H. M. and Fraser, J. P., "Deep Water Pipeline", API Paper No. 926-11-J, Houston, March 1966.

6. Miller, D. R., "Design and Construction of Submarine Pipelines", ASCE Transportation Eng. Conf., Minneapolis, May 17-21, 1965.

7. Ledford, R. C., "Design of Submarine Pipe Lines for Stability", Petroleum Engineer, May 1953, p. D-70.

8. "Determination of Net Buoyancy of Submerged Pipe Lines", Pipe Line Industry, Nov. 1963, p. 66.

9. Hausford, J. E. and Lubinski, A., "Effects of Drilling Vessel Pitch or Roll on Kelly and Drill Pipe Fatigue", Journal of Petroleum Technology, Jan. 1964.

10. Broussard, Barry and McCarron, "Effects of Variations in Laying System and Pipe Wall Thickness on Red Snapper Pipeline", Offshore, June 2, 1967, p. 47.

11. Stewart and Fraser, "Experimental Measurement of Stresses while Laying Pipe Offshore", ASME Paper 66-PET-24.

12. Kreign, J. L., "Good Engineering Practice Best Protection for Offshore Lines", ASCE, Vol. 91, No. PL-1, July 1965, p. 15, or Pipe Line Industry, March 1963, p. 43.

13. "How does Flexible Submarine Pipe Work for the Gas Industry", Gas, June 1964, p. 53.

14. Carina, P. L., "How Montubi Laid Red Sea Pipe Line", Pipe Line Industry, Dec. 1965, p. 37.

15. Alvim, J. P., "How to Design Pipelines for Negative Buoyancy", Pipeline Engineer, Dec. 1965, p. 48.

SELECTED REFERENCES

Pipe Stresses

16. Klohn, C. H., "How to Design Weight Coating with Computer", Pipeline Engineer, Sept. 1961, p. 284.

17. Lind, E. R., "How to Evaluate Pipe Stresses when Drilling from a Floating Vessel", World Oil, June 1961, p. 95.

18. Blumberg, R., "Hurricane Winds, Waves and Currents Test Marine Pipe Line Design", Pipe Line Industry, June 1964, p. 42; July 1964, p. 70; August 1964, p. 34; Sept. 1964; Oct.1964; November 1964, p. 85.

19. Ward, "Laying Large Diameter Offshore Pipelines", Offshore, June 2, 1967, p. 52.

20. Tesson, P. A., "Laying Pipe from a Reel", Offshore, July 1963, p. 33.

21. Ball, "Light-Media Pump for Dredging Minerals from the Sea Floor", ASME Paper No. 67-UNT-4, Underwater Tech. Conf., May 1967.

22. Sharifi, J. and Stierman, K. A., "Methods used in Laying Submarine Pipelines in the Persian Gulf", Le Petrole et la Mer, Section IV, No. 406.

23. Burnett, R. R., "Naphtha used to Submerge Submarine Pipeline", Petroleum Management, Oct. 1964, p. 167.

24. Riley, W. E., "New Approach to Pipeline Crossings", Oil and Gas Journal, June 12, 1961, p. 151.

25. Wittgenstein, F. W., "Novel Schemes for Laying Underwater Pipeline and for Carrying Oil and Gas in the Same Pipes", Pipes and Pipelines, Feb. 1962, p. 37.

26. Short, "Offshore Pipeline Anchoring System", Proceedings of OECON 1967, p. 603.

27. Bozeman, H. C., "Offshore Pipelining to go Deeper, Farther Out", Oil and Gas Journal, July 15, 1963, p. 119.

28. St. Denis and Armijo, "On the Dynamic Analysis of the Mohole Riser", Ocean Sci. and Ocean Eng., Vol. II, MTS, 1965, p. 1240.

29. Lampietti, F. J., "Pendulation of Pipes and Cables in Water", ASME Paper No. 63-WA-101, Journal of Eng. for Industry, August 1964, p. 299.

30. Short, "Pipeline Anchoring System", Offshore, June 2, 1967, p. 57.

31. Lynch, J. F., "Pipelines for the North Sea", Le Petrole et la Mer, Section IV, No. 404.

32. Minor, L. E., "Pipelines on the Continental Shelf", 2nd International Pipes, Pipelines, Pumps and Valves Convention, London, April 13-17, 1964.

SELECTED REFERENCES

Pipe Stresses

33. Bowman, R. L. and Bemis, J. H., "Pipeliners Whip Heavy Surf and Coral to Lay Offshore Hawaiian Lines", Pipe Line Industry, Jan. 1961, p. 33.

34. Lee and Bankston, "Pipelining Offshore", Offshore, June 2, 1967, p. 36.

35. Wilhoit and Merwin, "Pipe Stresses Induced in Laying Offshore Pipeline", ASME Paper 66-PET-5.

36. Holland, S. M., "Screw Anchors Hold Wandering Submarine Line", Pipe Line Industry, Dec. 1962, p. 54.

37. Berard, D. J. and Smith, E. S., "Significant Developments in Economical Underwater Hot-Tapping and Tie-In Equipment and Techniques", ASME Paper No. 65-PET-35.

38. Moyal, M., "Subsea Gasline", Gas Journal, May 8, 1963, p. 145.

39. Brueckner, "Underwater Drilling and Completion System", ASME Underwater Tech. Conf., May 1967.

40. Lamb, "Underwater Pipelines", Exploiting the Ocean, MTS, June 1966, p. 293.

41. "Unreeling Pipe from a Spool Cuts Offshore Laying Costs to only $2.00 per Foot", Pipe Line Industry, Jan. 1963, p. 29.

42. Berard, D. J., "Using Stingers in Offshore Pipelining", Pipe Line Industry, April, 1963, p. 43.

43. Dixon, D. A. and Rutledge, D. R., "Laying a Pipe Line in Deep Water under Tension - Without a Stinger", Ocean Industry, Dec. 1967, Vol. 2, No. 12, p. 32.

44. Garcia, D. J., Wilhoit, J. C. and Merwin, J. E., "Calculating Bending Moments in Laying Lines under Tension", Ocean Industry, Dec. 1967, Vol. 2, No. 12, p. 29.

45. Ward, D. R., "Submarine Pipeline Construction Techniques", Proceedings of OECON, 1966, p. 151.

46. Radbill, J. R., "Computation of Flow-Line Installation Stresses", Proceedings of OECON, 1968, p. 383.

SELECTED REFERENCES

Pressure Hull Structure

1. Navaratna, Pian and Witmer, "Analysis of Elastic Stability of Shells of Revolution by the Finite Element Method", AIAA/ASME Conf. March 29-31 1967, Palm Springs, California.

2. Vafakos, "Analysis of Uniform Deep Oval Reinforcing Rings", SNAME, Journal of Ship Research, Vol. 7, No. 4, April 1964, p. 21.

3. Stachiw, J. D., "Applicability of Concrete to Deep Submergence Hollow Shell Structures", Civil Eng. in the Oceans, ASCE, San Francisco, Sept. 6-8, 1967.

4. Wah, "Axisymmetric Buckling of Ring-Stiffened Cylindrical Shells under Axial Compression", SNAME, Journal of Ship Research, Vol. 9, No. 1, June 1965, p. 66.

5. Kendrick, "Buckling under External Pressure of Circular Cylindrical Shells with Evenly Spaced Equal Strength Circular Ring Frames", Part I, NCRE Report R 211.

6. Kendrick, "Buckling under External Pressure of Circular Cylindrical Shells with Evenly Spaced Equal Strength Circular Ring Frames", Part III, NCRE Report R 244.

7. Kendrick, "Buckling under External Pressure of Ring Stiffened Circular Cylinders", RINA, Vol. 107, 1965, p. 139.

8. Ross, "Collapse of Ring-Reinforced Cylinders under Uniform External Pressure", RINA, Vol. 107, 1965, p. 375.

9. Dunham and Heller, "Comparative Behavior of Submarine Pressure Hull Structures of Different Scales under Uniform External Pressure", ASNE Journal, May 1963, p. 397.

10. Rossi and Johnston, "Composite Sandwich for Small, Unmanned, Deep-Submergence Vehicles", ASME Paper 65-UnT-2.

11. Stachiw, "Concrete Pressure Hulls for Ocean Floor Installations", Journal of Ocean Technology, Vol. 1, No. 2, 1967, p. 19.

12. Kendrick, "Deformation under External Pressure of Nearly Circular Cylindrical Shells with Evenly Spaced Equal Strength Nearly Circular Ring Frames", NCRE Report R 259.

13. Stachiw and Snyder, "Design and Fabrication of Glass and Ceramic Deep Submergence, Free-Diving Instrumentation Capsules with Capabilities of 3500 Fathoms", ASME Paper 65-UnT-1.

14. Jones and Salerno, "Effect of Structural Damping on the Forced Vibrations of Cylindrical Sandwich Shells", ASME Paper 65-WA/Unt-1.

SELECTED REFERENCES

Pressure Hull Structure

15. Galletly and Bart, "Effects of Boundary Conditions and Initial Out-of-Roundness on the Strength of Thin-Walled Cylinders Subject to External Hydrostatic Pressure", ASME Paper 56-APM-9.

16. Ballow and Barry, "Elastic Instability of Relatively Thick Circular Cylindrical Shells Subjected to Hydrostatic Pressure", ASNE Journal, Aug. 1964, p. 621.

17. Midgley and Johnson, "Experimental Buckling of Internal Integral Ring-Stiffened Cylinders", Experimental Mechanics, April 1967, p. 145.

18. Wenk, "Feasibility of Pressure Hulls for Ultradeep Running Submarines", ASME Journal of Engineering for Industry, August 1962, p. 373.

19. Yamamato, "General Instability of a Reinforced Cylindrical Shell Clamped at Both Ends under External Pressure", Fourteenth Japan National Congress for Applied Mechanics, 1964.

20. Milligan and Lakshmikantham, "General Instability of Shallow-Stiffened Orthotropic Cylinders under Hydrostatic Pressure", SNAME, Journal of Ship Research, Vol. 11, No. 2, June 1967, p. 117.

21. Bryant, "Hydrostatic Pressure Buckling of a Ring-Stiffened Tube", NCRE Report R 306.

22. Healey, "Hydrostatic Tests of Two Prolate Spheroidal Shells", SNAME, Journal of Ship Research, Vol. 9, No. 2, Sept. 1965, p. 77.

23. Theocaris and Hill, "Inelastic Buckling of Rib-Cored Sandwich Cylinders under External Hydrostatic Pressure", ASME Paper 66-APM-5.

24. Block, "Influence of Prebuckling, Deformations, Ring Stiffeners and Load Eccentricity on the Buckling of Stiffened Cylinders", AIAA/ASME Conf. March 29-31, 1967, Palm Springs, California.

25. "Instability of Ring-Reinforced Cylindrical Shells under Uniform External Pressure", ASNE Journal, 1962, p. 31.

26. Ross, "Instability of Ring-Stiffened Circular Cylindrical Shells under Uniform External Pressure", RINA, Vol. 107, 1965, p. 157.

27. Nott, "Investigation on the Influence of Stiffener Size on the Buckling Pressures of Circular Cylindrical Shells under Hydrostatic Pressure", SNAME, Journal of Ship Research, Vol. 6, No. 2, Oct. 1962, p. 24.

28. Bert, "Large Weight Reductions Possible in Pressure Vessels", Space/Aeronautics, Oct. 1962, p. 77.

SELECTED REFERENCES

Pressure Hull Structure

29. Windenburg, "Master Charts for the Design of Vessels under External Pressure", ASME Trans. 1947, p. 345.

30. Gerard, "Minimum Weight Design of Ring Stiffened Cylinders under External Pressure", SNAME, Journal of Ship Research, Vol. 5, No. 2, Sept. 1961, p. 44.

31. Marbec, "Notes on the Collapsing of Curved Beams and Curved Elastic Strips", INA, Vol. 53, ii, p. 233, 1911.

32. Cook, "On Certain Approximations in Sandwich Plate Analysis", ASME Paper 66-APM-B.

33. Gerard and Lakshmikantham, "Optimum Thin-Wall Pressure Vessels of Anisotropic Materials", ASME Paper 66-APM-BB.

34. Lakshmikantham and Gerard, "Plastic Stability of Simply Supported Spherical Plates under External Pressure", SNAME, Journal of Ship Research, Vol. 8, No. 2, Sept. 1964, p. 1.

35. Gerard, "Plastic Stability Theory of Stiffened Cylinders under Hydrostatic Pressure", SNAME, Journal of Ship Research, Vol. 6, No. 2, Oct. 1962, p. 1.

36. Bernstein, "Pressure Hulls for Deep Submergence Vehicles", AIAA Journal of Hydronautics, 1967.

37. Becker, "Pressure Stability of Prolate Spheroids", SNAME, Journal of Ship Research, Vol. 11, No. 2, June 1967, p. 89.

38. Wenk, "Pressure Vessel Analysis of Submarine Hulls", Welding Research Supplement, June 1961, p. 272s.

39. Alexander, "Pressurized Membrane Container and its Applications", MTS, The New Thrust Seaward, p. 237.

40. Bodey, "Ring-Stiffener Parameters for Cylindrical Pressure Hulls with Circular Imperfections", SNAME, Marine Technology, Vol. 3, No. 4, Oct. 1966, p. 485.

41. Payne, "Some Observations with Regard to the Problem of the Structural Design of the Pressure Hull of Submarines", INA, Vol. 75, 1933, p. 237.

42. Whiting, "Strength of Submarine Vessels", INA, Vol. 63, 1921, p. 49.

43. Saunders & Windenburg, "Strength of Thin Cylindrical Shells under External Pressure", ASME Trans. 1931, p. 207.

44. Allentuch and Kempner, "Stresses in Eccentric Nonuniform Rings Reinforcing Cylindrical Shells", SNAME, Journal of Ship Research, Vol. 11, No. 2, June 1967, p. 73.

SELECTED REFERENCES

Pressure Hull Structure

45. Catudal, F. W. and Schneider, R. W., "Stresses in a Pressure Vessel with Circumferential Ring Stiffeners", The Welding Journal, December 1957, December 1957, p. 550-S.

46. Stromeyer, "Stresses in Stayed Cylindrical Shells", INA, Vol. 55, i, 1913, p. 201.

47. Vasta, Pohler, Becker and Winter, "Structural Problems in Penetrated Spheres under Pressure", ASNE Journal, Vol. 79, No. 2, April 1967, p. 207.

48. Vafakos, "Theorem on Buckling and Structural Stiffness", SNAME, Journal of Ship Research, Vol. 9, No. 1, June 1965, p. 9.

49. Vafakos and Restand, "Uniform Oval Reinforcing Rings", SNAME, Journal of Ship Research, Vol. 9, No. 2, Sept. 1965, p. 105.

50. Windenburg, "Vessels under External Pressure" Mechanical Engineering (ASME), August 1937, p. 601.

51. Jones, B. H., "Assessing Instability of Thin-Walled Tubes under Biaxial Stresses in the Plastic Range", Experimental Mechanics, Jan. 1968, p. 10.

52. Berkowits, A., Singer, J. and Weller, T., "Buckling of Unstiffened Conical Shells under Combined Loading", Experimental Mechanics, Nov. 1967, p. 458 and SESA Proceedings, Vol. XXIV, No. 2, p. 458.

53. Sute, W. T., "Design of Heavy-Wall Pressure Vessels", ASME Paper No. 67-PET-4.

54. Krenske, M., Jones, R. and Kiernan, T., "Design of Pressure Hulls for Small Submersibles", ASME Paper No. 67-WA/UnT-7.

55. Stachiw, J. D., "New Approach to the Assembly of Cylindrical Hulls from Shell Sections", Journal of Ocean Technology, Vol. 2, No. 1, Dec. 1967, p. 50 and Naval Engineers Journal, Vol. 80, No. 4, Aug. 1968, p. 590.

56. Brock, J. E., "Optimizing Design and Determining Stress Distributions of Multishell Cylindrical Pressure Vessels", ASME Paper No. 67-PET-32.

57. Krenzke, M. A., "Structural Aspects of Hydrospace Vehicles", Naval Engineers Journal, Vol. 77, No. 4, August 1965, p. 597.

58. MacNaught, D. F., "Submarine Pressure-Hull Design", Principles of Naval Architecture, Comstock, J. P., 1967, SNAME, Chapter IV, Section 8, p. 206.

59. Howell, J. D., "Thick-Walled Cylinder Construction for Deep Submergence", Critical Look at Marine Technology, MTS, 1968, p. 549.

SELECTED REFERENCES

Pressure Hull Structure

60. Evans, J. H., "Design of Stiffened and Unstiffened Cylindrical Shells under Hydrostatic Pressure", Lecture Notes, M.I.T. Department of Naval Architecture and Marine Engineering.

61. "Guide for the Classification of Manned Submersibles", American Bureau of Shipping, 1968.

62. Korkut, M. D., "Proposed Design Method for Cylindrical External Pressure Compartments", Proceedings of OECON, 1968, p. 667.

The following are reports of the U. S. Navy David Taylor Model Basin, Carderock, Maryland, (now Naval Ship Research and Development Center).

63. Krenzke, M., Hom, K. and Proffitt, J., "Potential Hull Structures for Rescue and Search Vehicles of the Deep-Submergence Systems Project", Report No. 1985, March 1965.

64. Krenzke, M. and Ward, G. D., "Feasibility of a Connected Sphere Pressure Hull for the Rescue Vehicle of the Deep-Submergence Systems Project", Report No. 2007, July 1965.

65. Zilliacus, S. and Hashmall, H., "Impact Strength and Response of Protected Brittle Models of Deep Submergence Structures", Report No. 2314, April 1967.

66. Couch, W. P., "Hydrostatic and Cyclic Tests of Four, Hollow-Filament, Glass-Reinforced Plastic Cylinders with Titanium Hemispherical End Closures", Report No. 2402, June 1967.

67. Macurdy, A. C., "Exploratory Investigation of Nonwelded Pressure Hulls for Hydrospace Vehicles", Report No. 1762, March 1964.

68. Healey, J. J., "Exploratory Tests of Cylinders with Various Lightweight Stiffening Systems under External Hydrostatic Pressure", Report No. 2073, August 1965.

69. Nott, J. A. and Ward, G. D., "Evaluation of Stresses in Web-Stiffened, Cylindrical Sandwich Shells Subjected to Uniform External Pressure", Report No. 2092, Sept. 1965.

70. Raetz, R. V., "Tests of Fabricated Multilayered Ring-Stiffened Cylindrical Models under External Hydrostatic Pressure", Report No. 2173, April 1966.

71. Fishlowitz, E. G., "Large-Scale Model Evaluation of a Welded, Web-Stiffened Titanium Sandwich Hull for Deep-Submergence Applications", Report No. 2412, July 1967.

72. Lunchick, M. E., "Plastic Buckling Pressure for Spherical Shells", Report No. 1493, July 1963.

SELECTED REFERENCES

Pressure Hull Structure

73. Krenzke, M. A. and Kiernan, T. J., "Effect of Initial Imperfections on the Collapse Strength of Spherical Shells", Report 1757, Feb. 1965.

74. Krenzke, M. A. and Charles, R. M., "The Elastic Buckling Strength of Spherical Glass Shells", Report No. 1759, Sept. 1963.

75. Healey, J. J., "Parametric Study of Unstiffened and Stiffened Prolate Spheroidal Shells under External Hydrostatic Pressure", Report No. 2018, Aug. 1965.

76. Hyman, B. I., "Elastic Instability of Prolate Spheroidal Shells under Uniform External Pressure", Report No. 2105, Dec. 1965.

77. Kloppel, K. and Jungbluth, O., "Contribution to the Durchschlag Problem of Thin-Walled Spherical Shells (Experiments and Design Formulas)", Translation No. 308, May 1966.

78. Savin, G. N., et al, "Spherical Shell Weakened by Two Unequal Circular Holes", Translation No. 320, March 1965.

79. von Mises, "Critical External Pressure of Cylindrical Tubes, The", Report No. 309.

80. "Draft of Proposed A.S.M.E. Boiler Construction Code", Report No. 329.

81. Windenburg, "Proposed Rules for Construction of Unfired Vessels Subjected to External Pressure", Report No. 356.

82. von Mises, "Critical External Pressure of Cylindrical Tubes under Uniform Radial and Axial Pressure", Report No. 366.

83. Windenburg and Trilling, "Collapse by Instability of Thin Cylindrical Shells under External Pressure", Report No. 385.

84. Windenburg, "Therotical and Empirical Equations Represented in Proposed Rules for the Construction of Unfired Pressure Vessels", Report No. 387,

85. Trilling, "Influence of Stiffening Rings on the Strength of Thin Cylindrical Shells under External Pressure", Report No 396.

86. Nash, W. A., "Buckling of Multiple-Bay Ring-Reinforced Cylindrical Shells Subject to Hydrostatic Pressure", Report 785, April 1954.

87. Kaminsky, "General Instability of Ring-Stiffened Cylinders with Clamped Ends under External Pressure by Kendrick's Method", Report No. 855.

SELECTED REFERENCES

Pressure Hull Structure

88. Galletly, G. D. and Bart, R., "Effects of Boundary Conditions and Initial Out-of-Roundness on the Strength of Thin-Walled Cylinders Subject to External Hydrostatic Pressure", Report No. 1066, Nov. 1957.

89. Wenk and Kennard, "Weakening Effect of Initial Tilt and Lateral Buckling of Ring Stiffeners on Cylindrical Pressure Vessels", Report No. 1073.

90. Reynolds, T. E., "A Graphical Method for Determining the General Instability Strength of Stiffened Cylindrical Shells", Report No. 1106, Sept. 1957.

91. Lunchick, M. E. and Short, R. D., "Behavior of Cylinders with Initial Shell Deflection", Report 1150, July 1957.

92. Galletly, G. D. and Reynolds, T. E., "A Simple Extension of Southwell's Method for Determining the Elastic General Instability Pressure of Ring-Stiffened Cylinders Subject to External Hydrostatic Pressure", Report No. 1191, Feb. 1958.

94. Lunchick, M. E., "Yield Failure of Stiffened Cylinders under Hydrostatic Pressure", Report No. 1291, January 1959.

95. Reynolds, T. E. and Blumenberg, W. F., "General Instability of Ring-Stiffened Cylindrical Shells Subject to External Hydrostatic Pressure", Report No. 1324, June 1959.

96. Krenzke, M. A., "Effect of Initial Deflections and Residual Welding Stresses on Elastic Behavior and Collapse Pressure of Stiffened Cylinders Subjected to External Hydrostatic Pressure", Report No. 1327, April 1960.

97. Krenzke, M. A. and Short, R. D., "Graphical Method for Determining Maximum Stresses in Ring-Stiffened Cylinders under External Hydrostatic Pressure", Report No. 1348, Oct. 1959.

98. Reynolds, T. E., "Inelastic Lobar Buckling of Cylindrical Shells under External Hydrostatic Pressure", Report No. 1392, August 1960.

99. Lunchick, M. E., "Plastic Axisymmetric Buckling of Ring-Stiffened Cylindrical Shells Fabricated from Strain-Hardening Materials and Subjected to External Hydrostatic Pressure, Report No. 1393, Jan. 1961.

100. Lunchick, M. E., "Graphical Methods for Determing the Plastic Shell-Buckling Pressures of Ring-Stiffened Cylinders Subjected to External Hydrostatic Pressure", Report No. 1437, March 1961.

101. Lunchick, M. E., "Plastic Prebuckling Stresses for ing-Stiffened Cylindrical Shells under External Pressure", Report No. 1448, Jan. 1961.

SELECTED REFERENCES

Pressure hull Structure

102. Pulos, J. G. and Salerno, V. L., "Axisymmetric Elastic Deformations and Stresses in a Ring-Stiffened, Perfectly Circular Cylindrical Shell under External Hydrostatic Poessure", Report No. 1497, Sept. 1961

103. Hom, K., "Elastic Stresses in Ring Frames of Imperfectly Circular Cylindrical Shells under External Pressure Loading", Report No. 1505, May 1962.

104. Keefe, R. F. and Short, R. D. Jr., "A Method for Eliminating the Effect of End Conditions on the Strength of Stiffened Cylindrical Shells", Report No. 1513, Sept. 1961.

105. Pulos, J. G., "Axisymmetric Elastic Deformations and Stresses in a Web-Stiffened Sandwich Cylinder under External Hydrostatic Pressure", Report No. 1543, Nov. 1961.

106. Raetz, R. V., "Analysis of Stresses in Axisymmetric Shell Structures Utilizing Toroidal Shells as Reinforcing Rings", Report No. 1569, Jan. 1962.

107. Ball, W. E. Jr., "Formulas and Curves for Determining the Elastic General-Instability Pressures of Ring-Stiffened Cylinders", Report No. 1570, Jan. 1962.

108. Lunchick, M. E., "Plastic General-Instability Pressure of Ring-Stiffened Cylindrical Shells", Report No. 1587, Sept. 1963.

109. Nott, J. A., "Investigation on the Influence of Stiffener Size on the Buckling Pressures of Circular Cylindrical Shells under Hydrostatic Pressure", Report No. 1600, Dec. 1961.

110. Reynolds, T. E., "Elastic Lobar Buckling of Ring-Supported Cylindrical Shells under Hydrostatic Pressure", Report 1614, Sept. 1962.

111. Pulos, J. G., "Structural Analysis and Design Considerations for Cylindrical Pressure Hulls", Report No. 1639, April 1963.

112. Nott, J. A., "Investigation on the Influence of Stiffener Size on the Buckling Pressures of Circular Cylindrical Shells under Hydrostatic Pressure", Report No. 1688, Jan. 1963.

113. Hom, K., "Axisymmetric Elastic Stresses in Circular Cylindrical Shells Stiffened by Internal Channel Sections and Subjected to Uniform External Pressure Loading", Report 1811, May 1964.

114. Nott, J. A., "Graphical Analysis for Maximum Stresses in Sandwich Cylinders under External Uniform Pressure", Report 1817, May 1964.

SELECTED REFERENCES

Pressure Hull Structure

115. Nott, J. A., "Axisymmetric Stresses in Orthotropic, Web-Stiffened Sandwich Cylinders Loaded with Uniform External Pressure", Report No. 1859, April 1966.

116. Short, R. D., "Effective Area of Ring Stiffeners for Axially Symmetric Shells", Report No. 1894, March 1964.

117. Short, R. D., "Membrane Design for Stiffened Cylindrical Shells under Uniform Pressure", Report No. 1898, April 1965.

118. Basdekas, N. L., "A Survey of Analytical Techniques for Determining the Static and Dynamic Strength of Pressure-Hull Shell Structures", Report No. 2208, September 1966.

120. Allnutt, R. B., "Relation between Testing and Performance of Structures for Deep Sea Vehicles", Report No. 2256, August 1966.

121. Reynolds, T. E., "A Summary of Submarine Structural Research. Part I - Conventional Hull Configurations, Chapter 5: Stresses in Pressurized Shells of Revolution", Report No. 2264, Dec. 1966.

122. Lomacky, O., "Elastic Stress Analysis in the End Bays of Stiffened Circular Cylindrical Shells under External Hydrostatic Pressure", Report No. 2287, December 1966.

123. von Sonden, K. and Tolke, F., "On Stability Problems in Thin Cylindrical Shells", Translation No. 33, Dec. 1949.

124. von Sanden and Gunther, "Strength of Cylindrical Shells Stiffened by Frames and Bulkheads under Uniform External Pressure on all Sides", Translation No. 38.

125. Shimanskiy, Y. A., "Structural Mechanics of Submarines. Part I: Practical Methods and Examples of Calculations for the Hull Strength of Submarines", Translation 305-1, Feb. 1964.

126. Salet, G., "A Method for Calculating the Buckling Pressure of Thin Shells of Revolution using Thuloup's General Theory of Elastic Buckling. Part I: The Thuloup General Theory of Elastic Buckling. Part II: Calculation of the Buckling Load of Thin Shells of Revolution under Uniform Normal Pressure", Translation No. 336, April 1967.

127. Krenzke, M. A., "Tests of Machined Deep Spherical Shells under External Hydrostatic Pressure", Report No. 1601, May 1962.

SELECTED REFERENCES

Pressure Hull Structure

128. Krenzke, M. A., "The Elastic Buckling Strength of Near-Perfect Deep Spherical Shells with Ideal Boundaries", Report No. 1713, July 1963.

129. Kiernan, T. J. and Nishida, K., "The Buckling Strength of Fabricated HY-80 Steel Spherical Shells", Report No. 1721, July 1966.

130. Krenzke, M. A. and Kiernan, T. J., "Tests of Stiffened and Unstiffened Machined Spherical Shells under External Hydrostatic Pressure", Report No. 1741, August 1963.

131. Kiernan, T. J., "Predictions of the Collapse Strength of Three HY-100 Steel Spherical Hulls Fabricated for the Oceanographic Research Vehicle ALVIN", Report No. 1792, March 1964.

132. Dadley, A. E., "Tests of Machined High-Strength Steel Spherical Shells Subjected to External Hydrostatic Pressure", Report No. 1854, August 1964.

133. Healey, J. J., "Exploratory Tests of Prolate Spheroidal Shells under External Hydrostatic Pressure", Report No. 1868, June 1965.

134. Nishida, K., "Tests of Machined Multilayer Spherical Shells with Clamped Boundaries under External Hydrostatic Pressure", Report No. 2012, August 1965.

135. Nishida, K., "Inelastic Buckling of Machined High Strength Hemispheres with Ideal Boundaries", Report No. 2090, Sept. 1965.

136. Schwartz, F. M., "Hydrostatic Tests of a High Strength Steel Internally Stiffened Hemisphere", Report No. 2302, Jan. 1967.

137. Costello, M. G. and Nishida, K., "The Inelastic Buckling Strength of Fabricated HY-80 Steel Hemispherical Shells", Report No. 2304, April 1967.

138. Windenburg and Trilling, "Test of an 8/10 Scale Model of the Pressure Hull of the Submarine MARLIN (SS205)", Report No. 473.

139. Windenburg, "Full Scale Test of Conning Tower and Escape Trunk for Grenadier Class Submarines (SS209-211)", Report No. 477.

140. Windenburg and Trilling, "Strain and Deflection Measurements on the Submarines TAMBOR (SS198) and TRITON (SS201) during Deep-Submergence Tests", Report No. 478.

141. Windenburg, "Hydrostatic Test of a 1/12-Scale Model Pressure-Hull Section of the SS285 Class of Submarines", Report No. 515.

SELECTED REFERENCES

Pressure Hull Structure

142. Windenburg, "Strain and Deflection Measurements on the Submarine BALAO (SS285) during Deep-Submergence Test", Report No. 519.

143. Lunchick, M. E. and Overby, J. A., "An Experimental Investigation of the Yield Strength of a Machined Ring-Stiffened Cylindrical Shell (Model BR-7M) under Hydrostatic Pressure", Report No. 1255, Nov. 1958.

144. Keefe, R. F. and Overby, J. A., "An Experimental Investigation of Effect of End Conditions on Strength of Stiffened Cylindrical Shells", Report No. 1326, Dec. 1959.

145. Hom, K. and Couch, W. P., "Hydrostatic Tests of Inelastic and Elastic Stability of Ring-Stiffened Cylindrical Shells Machined from Strain-Hardening Steel", Report No. 1501, Dec. 1961.

146. Blumenberg, W. F. and Reynolds, T. E., "Elastic General Instability of Ring-Stiffened Cylinders with Intermediate Heavy Frames under External Hydrostatic Pressure", Report No. 1588, Dec. 1961.

147. Couch, W. P. and Pulos, J. G., "Progress Report Experimental Stresses and Strains in a Ring-Stiffened Cylinder of Oval Cross-Section (Major-to-Minor Axis Ratio of 1.5)", Report No. 1726, March 1963.

148. Couch, W. P., "Hydrostatic Pressure Tests of a Ring-Stiffened Cylinder of Oval Cross Section (Major-to-Minor Axis Ratio of 1.5)", Report No. 1788, March 1964.

149. Blumenberg, W. F., "The Effect of Intermediate Heavy Frames on the Elastic General-Instability Strength of Ring-Stiffened Cylinders under External Hydrostatic Pressure", Report No. 1844, Feb. 1965.

150. Proffitt, J. L., "Cyclic Loading Studies of Two Composite Construction Models", Report No. 1853, August 1964.

151. Boichot, L. and Reynolds, T. E., "Inelastic Buckling Tests of Ring-Stiffened Cylinders under Hydrostatic Pressure", Report No. 1992, May 1965.

152. Galletly, G. D., "On the In-Vacuo Vibrations of Simply Supported, Ring-Stiffened Cylindrical Shells", Report No. 1195, Feb. 1958.

153. Leibowitz, R. C., "Effects of Damping on Modes of Vertical Vibration of Hull of USS THRESHER (SSN593)", Report No. 1384, March 1960.

154. Geers, T. L. and Junger, M. C., "Transient Response of a Damped Mechanical Oscillator attached to a Shock-Wave-Excited, Submerged Cylindrical Shell", Report No. 2142, March 1966.

SELECTED REFERENCES

Pressure Hull Structure

155. Kennard, "Tripping of T-Shaped Stiffening Rings on Cylinders under External Pressure", Report No. 1079.

156. Krenzke, M.A. and Ward, G. D., "Feasibility of a Connected Sphere Pressure Hull for the Rescue Vehicle of the Deep-Submergence Systems Project", Report No. 2007, July 1965.

SELECTED REFERENCES

Reinforcements, Pressure Fittings and Structural Details

A. Pressure Vessels

1. Stachiw, "Conical Acrylic Windows for Deep Sea Applications", Undersea Technology, May 1966, p. 42.

2. Stachiw, "Critical Pressure of Conical Acrylic Windows under Short-Term Hydrostatic Loading", ASME Paper 66-WA/UnT-2.

3. Van Dyke, P., "Effects of Single Curvature on Stresses at Reinforced Circular Holes", AIAA/SNAME Advanced Marine Vehicles Meeting, Norfolk, Va., May 1967.

4. Haworth, "Electrical Cabling System for Star III Vehicle", ASME Paper 66-WA/UnT-11.

5. Dunham, "Fatigue Testing of Large-Scale Models of Submarine Structural Details", Marine Technology, Vol. 2, No. 3, July 1965, p. 299.

6. Stachiw, J. D., "Flat Acrylic Windows for Deep Sea Applications", Undersea Technology, July 1967, p. 23.

7. Bialkowski and Stachiw, "Metallic Seals for Deep Submergence Underwater Vehicles", Marine Technology, Vol. 2, No. 2, April 1965, p. 116.

8. Mavor, "Observation Windows of the Deep Submersible 'Alvin'", Journal of Ocean Technology, Vol. 1, No. 1, June 27, 1966, p. 2.

9. Fox, "Safety Factors and Defects in Bolted Joints in Submarine Sea Water Systems", ASNE Journal, 1965, p. 723.

10. Haworth and Regan, "Watertight Electrical Cable Penetrations for Submersibles - Past and Present", ASME Paper 65-WA/UnT-12.

11. Stachiw, J. D., "Critical Pressure of Flat Acrylic Windows under Short-Term Hydrostatic Loading", ASME Paper 67-WA/UnT-1.

12. Vasta, J., Pohler, C., Becker, H. and Winter, R., "Structural Problems in Penetrated Spheres under Pressure", Naval Engineers Journal, Vol. 79, No. 2, Apr. 1967, p. 207.

13. Snoey, M. R. and Stachiw, J. D., "Windows and Transparent Hulls for Man in Hydrospace", Critical Look at Marine Technology, MTS, 1968, p. 419.

SELECTED REFERENCES

Reinforcements, Pressure Fittings and Structural Details

A. Pressure Vessels

14. Stachiw, J. D., "Critical Pressure of Spherical Shell Acrylic Windows under Short-Term Pressure Loading", ASME Paper No. 68-WA/UNT-1.

15. Stachiw, J. D., "Influence of Joint Strength and its Location between Cellular Sandwich Shell Facings", MTS, _Journal of Ocean Technology_, Vol. 2, No. 4, Oct. 1968, p. 132.

16. Woodland, B. T., "Structures for Deep Submergence", _Space/Aeronautics_, March 1967, p. 100.

The following are reports of the U. S. Navy David Taylor Model Basin, Carderock, Md. (now Naval Ship Research and Development Center)

17. Krenzke, M. A., "Hydrostatic Tests of Conical Reducers between Cylinders with and without Stiffeners at the Cone-Cylinder Junctures", Report No. 1187, Feb. 1959.

18. Raetz, R. V., "Experimental Investigation of the Strength of Small-Scale Conical Reducer Sections between Cylindrical Shells under External Hydrostatic Pressure", Report No. 1397, March 1960.

19. Kiernan, T.J. and Krenzke, M.A., "An Experimental Investigation of Closures and Penetrations for Pressure Vessels of Composite Construction", Report No. 1732, Feb. 1964.

20. Nott, James A., "Hydrostatic Tests to Determine Stability and Stress Characteristics of Stiffened Cylinders with Quasi-Tube Inserts", Report No. 1874, Nov. 1964.

21. Proffitt, J. L., "Experimental Investigation of Penetrations and Closures for Glass-Reinforced-Plastic Pressure Hulls with Titanium Jackets", Report No. 2091, Sept. 1965.

22. Raetz, R. V. and Pulos, J. G., "A Procedure for Computing Stresses in a Conical Shell Near Ring Stiffeners or Reinforced Intersections", Report No. 1015, April 1958.

23. Short, R. D., and Bart, R., "Analysis for Determining Stresses in Stiffened Cylindrical Shells Near Structural Discontinuities", Report No. 1065, June 1959.

24. Raetz, R. V., "Analysis of Stresses at Junctures of Axisymmetric Shells with Flexible Insert Rings of Linearly Varying Thickness", Report No. 1444, Jan. 1961.

25. Lomacky, Oles, "A Summary of Submarine Structural Research. Part 1: Conventional Hull Configurations. Chapter IX - Hull Penetrations", Report No. 2309, May 1967.

SELECTED REFERENCES

Reinforcements, Pressure Fittings and Structural Details

A. Pressure Vessels

26. Savin, G. N., et al, "Spherical Shell Weakened by Two Unequal Circular Holes", Translation No. 320, March 1965.

27. Krenzke, M. A. and Ward, G. D., "Feasibility of a Connected Sphere Pressure Hull for the Rescue Vehicle of the Deep-Submergence Systems Project", Report No. 2007, July 1965.

28. Lomacky, Oles, "Stress Concentration Factors for Small Reinforced Circular Penetrations in Pressurized Cylindrical Shells", Report No. 2559, Oct. 1967.

B. Tubular Structures

1. Bouwkamp, J. G., "Concept of Tubular Joint Design", Journal of the Structural Div., ASCE, April 1964.

2. Bouwkamp, J. G., "Considerations in the Design of Large-Size Welded Tubular Truss Joints", Engineering Journal, AISC, Vol. 2, July 1965, p. 88.

3. Bouwkamp, J. G., "Design of Tubular Joints with Gusset Plates", Civil Eng. in the Oceans, ASCE, San Francisco, Sept. 6-8, 1967, p. 241.

4. Bouwkamp, J. G., "Research on Tubular Connections in Structural Work", Welding Research Council Bulletin No. 71.

5. Bjilaard, P. P., "Stresses from Local Loadings in Cylindrical Pressure Vessels", ASME Paper No. 54-PET-7.

6. Bjilaard, P. P., "Stresses from Radial Loads in Cylindrical Pressure Vessels", Welding Research Supplement, 1954, p. 615s.

7. Bouwkamp, J. G., "Tubular Joint Design", AISC Engineering Journal, July 1965, p. 88.

8. Bouwkamp, J. G., "Tubular Joints under Slow-Cycle Alternating Loads", RILEM Symposium, Mexico City, Sept. 1966.

9. Bouwkamp, J. G., "Tubular Joints under Static and Alternating Loads", College of Engineering, University of California, Berkeley, SEL Report No. 66-15, June 1966.

10. Johnston, L. P., "Welded Tubular Joint Problem in Offshore Oil Structures", Soc. of Petroleum Engineers of AIME Paper No. SPE484.

11. Ward, D. R., "Rehabilitation Method Doubles Platform Life", Offshore, Vol. 28, No. 12, Nov. 1968, p. 48.

SELECTED REFERENCES

Salvage and Emplacement Operations

1. Miller and Webster, "Analysis of Large Object Lift Systems", ASME Under-
    water Tech. Conf., May 1967.

2. "Collapsible Salvage Pontoons", NavShip Systems Command Technical News,
    January 1967, p. 18.

3. Quirk, "Deep-Ocean Reactor Placement Study", ASME Paper 65-WA/UnT-11.

4. Land, E. S., "Development in Ground Tackle for Naval Ships", SNAME,
    Vol. 42, 1934, p. 164.

5. Keays and Harrington and Wardwell, "Diving and Salvage Operations on the
    Continental Shelf", Man's Extension into the Sea, MTS, January 1966, p. 176.

6. Winer and Numger, "Foam-in-Salvage", Nav. Eng. Journal, Vol. 79, No. 3,
    June 1967, p. 465.

7. Sullivan, "Marine Salvage", SNAME, Vol. 56, 1948, p. 104.

8. "New Salvage Lift System", NavShip Systems Command Technical News,
    January 1967, p. 23.

9. Bacon, "Notes on the Causes of Accidents to Submarine Boats and their
    Salvage", INA, Vol. 47, ii, 1965, p. 406.

10. Holm, C. H., "Practical Aspects of Ship Salvage, Rescue and Recovery",
    Ocean Industry, Vol. 2, No. 7, July 1967, p. 25.

11. Critchley, "Salvage of H.M.S. 'Thetis'", INA, Vol. 82, 1940, p. 109.

12. "Submersible Chamber Working Off Surinan", Undersea Technology, Sept.
    1966, p. 37.

13. Tucker, A. J., "What Bag Pontoons can do in Marine Construction", Pipe
    Line Industry, August 1963, p. 12.

14. Keays, K., "Casino Royale - British Royal Navy Submarine Salvage Exercise",
    Critical Look at Marine Technology, MTS, 1968, p. 103.

15. Holm, C. H., "Factors Involved in Raising a Ship", Ocean Industry, Vol. 2,
    No. 11, Nov. 1967, p. 26.

16. Ulrich, J. L., Mott, G. D. and Keyser, D. R., "How to Minimize Barge
    Damage During Salvage", Ocean Industry, Vol. 3, No. 7, July 1968, p. 78.

SELECTED REFERENCES

Salvage and Emplacement Operations

17. Davidson, W. M. and Stewart, R. L., "Modular Recovery Systems", _Critical Look at Marine Technology_, MTS, 1968, p. 141.

18. Dicus, W. A., "New Concept in Deep Water Salvage", _Critical Look at Marine Technology_, MTS, 1968, p. 87.

19. Holm, C. H., "Sweepwire Method of Finding Objects in Deep Water", _Ocean Industry_, Vol. 2, No. 10, Oct. 1967, p. 50.

20. Hunley, W. H., "Deep-Ocean Work Systems", _Ocean Engineering_, John Wiley and Sons, Inc. (Chapter 14) p. 493.

SELECTED REFERENCES

Submarine Vehicles (General)

1. "Aluminaut", ASNE Journal, 1965, p. 186 or Undersea Technology, Sept. 1964.

2. Loughman, R. R., "Aluminaut Tests and Trials", Ocean Sci. and Ocean Eng., Vol. II, MTS, 1965, p. 876.

3. "ALVIN - Ocean Research Submarine", ASNE Journal, Dec. 1964, p. 963.

4. Beran, W. T. and Rosencrantz, D. M., "Asherah Design and Operations", ASME Paper 65-UnT-3.

5. Sims, "British Submarine Design During the War", INA, Vol. 89, 1947, p. 149.

6. Horton, T. F., "Deep Diving Manned Submersibles", Exploiting the Ocean (Supplement), MTS, June 1966, p. 29.

7. Horman, "Deep Jeep", ASNE Journal, Feb. 1967, p. 145, or Undersea Technology, July 1966, p. 33.

8. Daubin, S. C., "Deep Ocean Work Boat (DOWB)", AIAA/SNAME, May 1967, Paper No. 67-370.

9. "Deep-Sea Vehicles' Design Considerations", ASNE Journal 1966, p. 515.

10. "Deep Submergence Vehicles", NavShip Systems Command Technical News, Jan. 1967, p. 56.

11. Berman, "Design and Analysis of Commercial Pressure Vessels to 500,000 psi", ASME Paper 65-WA/PT-1.

12. Eliot, F., "Design and Construction of the Deepstar 2000", The New Thrust Seaward, MTS, 1967, p. 479.

13. Jensen and Walsh, "Design and Operation of the Bathyscaphe 'Trieste'", SNAME, S. California, Nov. 21, 1960.

14. Oakley, O. H., "Design Considerations for Manned Deep-Sea Vehicles", ASNE Journal, June 1966, p. 515.

15. Feldman and Cathers, "Design of the Deep Submergence Rescue Vehicle", Ocean Sci. and Ocean Eng., Vol. I, MTS, 1965, p. 641.

16. Leiss, Whitmarsh and Wellington, "DIVEAR - An Unmanned Acoustics/ Oceanographic Research Vehicle", Ocean Sci. and Ocean Eng., Vol. II, MTS, 1965, p. 864.

17. Harvey, "Economics of Pressure Vessel Design", ASNE Journal, 1965, p. 51.

18. "Evolution of the Attack Submarine", ASNE Journal, 1962, p. 425.

SELECTED REFERENCES

Submarine Vehicles (General)

19. Starks, "German 'U'-Boat Design and Production", INA, Vol. 90, 1948, p. 291

20. Schade, "German Wartime Technical Developments", SNAME, Vol. 54, 1946, p. 83.

21. Moore, "Industrial Approach to Manned Submersible Diving Operations", ASME Underwater Tech. Conf., May 1967.

22. Fitch, K. R. and Munz, R. J., "Manned Submersible Development", AIAA/SNAME, Paper No. 67-372, May 1967.

23. Pritzlaff, J. A. and Munske, R. E., "Manned Submersibles of the World", Undersea Technology, August 1964, p. 20, or ASNE Journal, 1965, p. 715.

24. Clark, "Methods and Techniques for Sea-Floor Tasks", Ocean Sci. and Ocean Eng., Vol. I, MTS, 1965, p. 267.

25. Groves, "Mini-Subs", ASNE Journal, Vol. 79, No. 2, April 1967, p. 249.

26. Arentzen and Mandel, "Naval Architectural Aspects of Submarine Design", SNAME, Vol. 68, 1960, p. 622.

27. Reidy, "Ocean-Bottom Survey Vehicle", ASME Underwater Tech. Conf., May 1967.

28. Kissinger, Wenk, DeHart, and Mandel, "Oceanographic Research Submarine of Aluminum for Operation to 15,000 ft.", RINA, Vol. 102, 1960, p. 555.

29. "Offshore Drilling Practices", World Oil, May 1965, p. 110.

30. Heller, "Personal Philosophy of Structural Design of Submarine Pressure Hulls", ASNE Journal, 1962, p. 223.

31. Vincent and Stavovy, "Promising Aspects of Deep-Sea Vehicles", SNAME, Vol. 71, 1963, p. 220.

32. Peach, "Purposes and Method of Achieving a Deep Diving Submarine", ASNE Journal, 1963, p. 575.

33. McKee, "Recent Submarine Design Practices and Problems", SNAME, Vol. 67, 1959, p. 623.

34. Higham, "Royal Navy's Freak Submarine Designs", ASNE Journal, 1959, p. 63.

35. Braucart and Hoffman, "Star II, A Second Generation Research Submarine", MTS, The New Thrust Seaward, p. 459.

SELECTED REFERENCES

Submarine Vehicles (General)

36. Toher, R. A. and Dawson, J. H., "Star III - Design and Construction",
     ASME Paper 66-WA/UnT-8.

37. Krenzke, "Structural Aspects of Hydrospace Vehicles", ASNE Journal,
     1965, p. 597.

38. Alsager, "Submarine Design - A Multitude of Complex Problems",
     ASNE Journal, August 1961, p. 539.

39. Archbold, "Submarine Design for Work and Research, Beaver Mark IV,"
     AIAA/SNAME, May 1967, Paper No. 67-371.

40. Strasburg, "Submarine for Research in Fisheries and Oceanography",
     Ocean Sci. and Ocean Eng., Vol. I, MTS, 1965, p. 568.

41. Crewe and Hardy, "Submarine Ore Carrier", RINA, Vol. 104, 1962, p. 393.

42. Russo, Turner and Wood, "Submarine Tankers", SNAME, Vol. 68, 1960, p. 692.

43. "Swimmer Delivery Vehicles", NavShip Systems Command Technical News,
     Jan. 1967, p. 11.

44. Shumaker, L. A., "Trieste II", ASNE Journal, August 1964, p. 513.

45. "Undersea Vehicles for Oceanography", U. S. Gov't. Publ. ICO Pamphlet
     No. 18, Oct. 1965.

46. Harmstarf and McBride, "Underwater Apparatus for Deep Embedment of Cables
     and Pipes in Submarine Subsoil", Ocean Sci. and Ocean Eng., Vol. II,
     MTS, 1965, p. 657.

47. Bodey, C. E., "Undersea Machine", Mechanical Engineering, June 1965,
     Vol. 87, No. 6, p. 37.

48. Murphy and Nodland, "Unmanned Research Vehicle for Use Down to Mid-Ocean
     Depths", Ocean Sci. and Ocean Eng., Vol. II, MTS, 1965, p. 898.

49. Heller, R. K., "Accomplishments of the Cable Controlled Underwater
     Research Vehicle", AIAA Symposium on Modern Development in Marine
     Engineering, April 1966, p. 169.

50. Rainnie, W. O., Jr., "Adventures of ALVIN", Ocean Industry, Vol. 3, No. 5,
     May 1968, p. 22.

51. Walsh, J. B. and Rainnie, W. O., Jr., "ALVIN, An Oceanographic
     Research Submarine", ASME Paper No. 63-WA-160.

SELECTED REFERENCES

Submarine Vehicles (General)

52. Mavor, J. W., Froehlich, H. E., Marquet, W. M. and Rainnie, W. O., "ALVIN, 6,000 Ft. Submergence Research Vehicle", SNAME, Vol 74, 1966, p. 106.

53. Conway, J. C., "Benthos - Deep Submergence Pyroceram Test Vehicle", ASME Paper No. 67-DE-5.

54. Gray, G. M., "Capability and Cost Interactions in Deep Submersible Vehicles", Critical Look at Marine Technology, MTS, 1968, p. 485.

55. Linderoth, L. S., "Challenge of Designing Undersea Devices, The", ASME Paper No. 67-DE-29.

56. Archbald, F. G., "Characteristics of Beaver Mark IV Hull", Ocean Industry, Vol. 2, No. 10, Oct. 1967, p. 42.

57. Forman, W. R., "Deep Jeep - From Design through Operation", Journal of Ocean Technology, Vol. 2, No. 3, July 1968, p. 17.

58. Terry, R. D., "Deep Submersible", Western Periodicals Co.

59. Snyder, R. F., "Divear II - A High-Strength Aluminum Submersible", ASME Paper No. 67-DE-6.

60. "Efficient British Research Sub", Oceanology International, Mar/April 1968, p. 18.

61. Strum, R. G. and Kaarsburg, E. A., "Exploritory Submersible for the Future", Ocean Industry, Vol. 3, No. 8, August 1968, p. 50.

62. Willm, P. H., "French Bathyscaph Program", Third U.S.N. Hydrodynamics Symposium, p. 475.

63. Terry, J., "Lockheed's Deep Quest for 7,000 lb., 8,000 ft. Operating Depth", Ocean Industry, Vol. 2, No. 10, Oct. 1967, p. 45.

64. Wenzel, J. G. and Helvey, W. M., "Manned Aspects of Deep Submersibles", Man's Extension into the Sea, MTS, 1966, p. 111.

65. "North American Develops Concept of Miniature Diver Transport Sub", Ocean Industry, Vol. 2, No. 10, Oct. 1967, p. 61.

66. Hall, J. B., "Potential Utility of Research Submarines in the Offshore Petroleum Industry", ASME Paper 67-PET-33.

67. "Russia Converts Sub for Fishing Research", Ocean Industry, Vol. 3, No. 6, June 1968, p. 40.

SELECTED REFERENCES

Submarine Vehicles (General)

68. Linderoth, L. S., Jr., "Small Sub: Part I - Design Challenges", Mechanical Engineering, Vol. 90, No. 6, June 1968, p. 28.

     "Small Sub: Part II - Boon to Industry", Mechanical Engineering, Vol. 90, No. 7, July 1968, p. 22.

70. Loeser, H. T. and Dawson, J. H., "Star I - Design and Operation", Naval Engineers Journal, Vol. 77, No. 4, August 1965, p. 587.

71. Wenk, E., Jr., "Submarines to Advance Study and Effective Use of the Sea", Experimental Mechanics, Vol. 8, No. 8, August 1968, p. 337.

72. "Submersible Work Vessels Now Operating World-Wide", Ocean Industry, Vol. 3, No. 2, Feb. 1968, p. 4.

73. Pritzlaff, J. A., "Submersible Safety, Classification, Certification and the Law", ASME Paper No. 67-WA/UnT-5.

74. Luckow, W. K., "Trade-off Studies of a Deep Ocean Search Submersible", Critical Look at Marine Technology, MTS, 1968, p. 163.

75. Pritzlaff, J. A., "Underwater Work and Manned Submersibles", SAE Paper 670183.

76. Leary, F., "Anti-Sub Submarine", Space/Aeronautics, July 1966, p. 52.

77. Hunley, W. H., "Deep-Ocean Work Systems", Ocean Engineering, John Wiley and Sons, Inc. (Chapter 14) p. 493.

78. Busby, R. F., "Design and Operational Performance of Manned Submersibles", ASME Paper No. 68-WA/UNT-11.

79. Perry, J. H., "Design Approach to Lock-Out Submarines", Southeast Section S.N.A.M.E., May 10, 1968.

80. Pritzlaff, J. A., "Design Characteristics of the Deepstar Family of Vehicles", ASME Paper No. 68-WA/UNT-3.

81. Combs, G. A., "Design of Wet Submersibles", Southeast Section S.N.A.M.E., May 10, 1968.

82. Busby, R. F., Hunt, L. M. and Rainnie, W. O., "Hazards of the Deep", Ocean Industry, Part I, Vol. 3, No. 7, July 1968, p. 72; Ocean Industry, Part II, Vol. 3, No. 8, Aug. 1968, p. 32; Ocean Industry, Part III, Vol. 3, No. 9, Sept. 1968, p. 53.

83. Pritzlaff, J. A. and Shenton, E. H., "Manned Submersible - A Way to do Undersea Work", Proceedings of OECON, 1966, p. 543.

SELECTED REFERENCES

Submarine Vehicles (General)

84. Bodey, C. E. and Friedland, N., "Naval Architecture of Submarine Work Boats for Offshore Work", _Proceedings of OECON_, 1966, p. 105.

85. Rainnie, W. O. Jr., "Operating Season with ALVIN for Science", MTS, _Journal of Ocean Technology_, Vol. 2, No. 4, Oct. 1968, p. 11.

86. Luckow, W. K., "Subsystems Comparisons and System Sensitivities for a Deep Ocean Work Boat", _Proceedings of OECON_, 1968, p. 733.

87. Oakley, O. H., "Vehicles and Mobile Structures", _Ocean Engineering_, John Wiley and Sons, Inc. (Chapter 11) p. 350.

88. Leary, F., "Vehicles for Deep Submergence", _Space/Aeronautics_, April 1968, p. 52.

SELECTED REFERENCES

Subsurface
Habitations, Observatories and Production Units

1. Spies, "Comprehensive Life Support System for the Makai Range Sea Floor Laboratory and Ocean Test Facility", MTS, The New Thrust Seaward, p. 315.

2. Burkart, "Deep Ocean Engineering at BuDocks", Ocean Sci. and Ocean Eng., Vol. I, MTS, 1965, p. 171.

3. McKinnon, "Design, Construction and Outfitting of SEALAB II", Man's Extension into the Sea, MTS, January 1966, p. 14.

4. McKinnon, "Design, Construction and Outfitting of SEALAB II", ASNE Journal, 1966, p. 179.

5. Clark, "Engineering Concept of a Sea-Floor Laboratory", MTS, The New Thrust Seaward, p. 543.

6. Austin, C. F., "Manned Undersea Installation and Civil Engineers", Civil Eng. in the Oceans, ASCE, San Francisco, Sept. 6-8, 1967.

7. Dugan, "Manned Undersea Stations", Ocean Sci. and Ocean Eng., Vol. I, MTS, 1965, p. 652.

8. Wolfe, "Manned Chamber Design for Prolonged Submergence", ASME Underwater Tech. Conf., May 1967.

9. "Nemo -A New Undersea Observatory", Undersea Technology, June 1966, p. 39.

10. Blockwick, "Ocean Engineering Aspects of SEALAB II", ASNE Journal, 1966, p. 609.

11. Hromadik, "Ocean Engineering for Human Exploration", Man's Extension into the Sea, MTS, January 1966, p. 74.

12. O'Neal, "Overview of SEALAB II", Man's Extension into the Sea, MTS, January 1966, p. 4.

13. Link, "Pressure Equalized Flexible Structures for Manned Dwellings", Man's Extension into the Sea, MTS, January 1966, p. 207.

14. "SEALAB I", ASNE Journal, February 1966, p. 35.

15. McKinnon, "SEALAB II", Marine Technology, Vol. 4, No. 1, January 1967, p. 591.

16. "SEALAB III", NavShips Tech. News, July 1967, p. 13.

17. Jue, "Some Naval Architectural Aspects of SEALAB II", U. S. N. Assoc. of Senior Eng., March 31, 1967.

SELECTED REFERENCES

Subsurface Habitations, Observatories and Production Units

18. Hakkarinen, "World of NOMAD-1", Buoy Technology, MTS, March 1964, p. 443.

19. Sparkman, J. C. and Foster, W. E., "California Company's New Mobile 20,000 bbl. Underwater Storage Unit", API, DPP, 1961, p. 114.

20. Rebicoff, D. I., "Case for Unmanned Undersea Systems", Sea Frontiers, Vol. 13, No. 3, May-June, 1963.

21. "Dubai Tank Drydock Project Underway", Offshore, Vol. 28, No. 9, August 1968, p. 145.

22. "Government and Industry to Team Up for Record 60-Day Undersea Mission", Ocean Industry, Vol. 3, No. 7, July 1968, p. 35.

23. "Ocean Drilling and Exploration Slashes Costs with Underwater Oil Storage", Offshore, July 1960, p. 11.

24. Hanna, F. K., "Underwater Fuel and Oil Storage", AIME Soc. of Petroleum Engineers, Paper SPE730.

25. "Underwater Oil Storage", Undersea Technology, Jan. 1964, p. 50.

26. "Underwater Storage System", Offshore, Sept. 1960, p. 10.

27. Eager, W. J., "Sealab III Complex", Undersea Technology, Vol. 9, No. 8, August 1968, p. 28.

28. Danforth, L. J., "Feasibility of an Offshore Underwater Oil-Drilling Platform", Proceedings of OECON, 1966, p. 159.

29. Culpepper, W. B., Frost, W. P. and Porter, R. B., "Desig of Underseas Habitats", Southeast Section S.N.A.M.E., May 10, 1968.

30. "Mid-Water Production Capsules to Work Shallow Oil Sands", Ocean Industry, Vol. 3, No. 10, Oct. 1968, p. 71.

31. Gallery, P. D., "Sealab III - Giant Step Toward Occupation of the Sea Floor", Ocean Industry, Vol. 3, No. 10, Oct. 1968, p. 65.

32. "OSI Unveils Underwater Production System", Offshore, Vol. 28, No. 11, October 1968, p. 83.

SELECTED REFERENCES

Surface Observatories and Instrument Platforms

1. Faires and Pida, "FORDS (Floating Ocean Research and Development Station)", Buoy Technology, MTS, March 1964, p. 250.

2. Rudnick, "FLIP (Floating Instrument Platform)", Buoy Technology, MTS, March 1964, p. 502.

3. Picard, "One Thousand Ton Meterological Buoy", M.I.T. S.M. Thesis (XIII) June 1964.

4. Talley, "SPAR", Buoy Technology, MTS, March 1964, p. 269.

5. Fisher, F. H. and Spiess, F. N., "FLIP - Floating Instrument Platform", Journal of the Acoustical Soc. of America, Vol. 35, No. 10, Oct. 1963, p. 1633.

6. Holmes, J. F., "Oceanographic Research Ship", Critical Look at Marine Technology, MTS, 1968, p. 529.

7. Glosten, L. R., "Ocean Platforms - With Particular Interest to Scripps Institute's FLIP", Trans. Inst. of Marine Eng. - Canadian Div., No. 19, March 1965, p. 1.

8. Glosten, L. R., "FLIP - Some Remarks on Certain Design Considerations", SNAME, Pacific Northwest Sect., 6 Oct. 1962.

9. Rudnick, P., "FLIP: An Oceanographic Buoy", Science, No. 3649, Vol. 146, 4 Dec. 1964, p. 1268.

10. Kaufman, R., "Design Features of a Stable Platform for Acoustic Research - SPAR", SNAME, Philadelphia Section, 13 Dec. 1965.

11. Marks, W. and Jarlan, G. E., "Some Unique Characteristics of a Perforated Cylindrical Platform for Deep-Sea Operations", Proceedings of OECON, 1966, p. 349.

12. Spiess, F. N., "Oceanographic and Experimental Platforms", Ocean Engineering, (Brahtz), John Wiley and Sons, Inc. (Chapter 15) p. 553.

13. Marks, W. and Jarlan, G. E., "Some Unique Characteristics of a Perforated Cylindrical Platform for Deep-Sea Operations", Proceedings of OECON, 1966, p. 349.

14. Holmes, J. F., "Vertically Stable Floating Drill Platform", Proceedings of OECON, 1968, p. 7.

SELECTED REFERENCES

Surface Support Vessels

1. Silverman and Gaul, "Concept of Portability Applied to Future Oceanographic Ship Operation", Ocean Sci. and Ocean Eng., Vol. I, MTS, 1965,
   p. 384.

2. Parham, "Deep Quest Support Operations", Proceedings of OECON 1967, p. 359.

3. Goodier, "Marine Mining Research Ships", Proceedings of OECON 1967, p. 627.

4. Andrews, J. N., Pincus, D. S., Dinsenbacher, A. L., "Model Test Determination of Sea Loads on Catamaran Cross Structure", May 1967,
   (Unclassified), Naval Ship Research and Development Center, Report 2378.

5. Stine, W. D., "Offshore Drilling Tenders", API, DPP, 1954, p. 267.

6. Lankford, B. W., "Structural Design of the ASR Catamaran Cross-Structure",
   Nav. Eng. Journal, Vol. 79, No. 4, August 1967, p. 625.

7. Meier, H. A., "Preliminary Design of a Catamaran Submarine Rescue Ship",
   Marine Technology, Vol. 5, No. 1, January 1968, p. 72.

8. Hamlin, "Catamaran as a Seagoing Work Platform", Ocean Sci. and Ocean
   Eng., Vol. II, MTS 1965, p. 1127.

SELECTED REFERENCES

Testing Facilities and Experimental Mechanics

1. Osgerby, "Application of the Moiré Method for use with Cylindrical Surfaces", Experimental Mechanics, July 1967, p. 313.

2. Carlson, Sendelbeck and Haff, "Experimental Studies of the Buckling of Complete Spherical Shells", Experimental Mechanics, July 1967, p. 281.

3. Gray, "Laboratory Simulation of the Deep-Ocean Environment", AIAA Journal of Hydronautics, 1967.

4. Svarez and Kramarow, "Large Prestressed Concrete Vessels for Deep Submergence Testing", ASME Paper 65-WA/UnT-8.

5. Theocaris, "Moiré Topography of Curved Surfaces", Experimental Mechanics, July 1967, p. 289.

6. Johnston and Imbach, "Deep Ocean Environmental Laboratory", Ocean Sci. and Ocean Eng., Vol. II, MTS 1965, p. 1205.

7. Keller, K. H., "High Pressure Test Chambers - State-of-the-Art", ASME Paper No. 68-WA/UNT-8.

8. Stachiw, J. D., "Hydrospace - Environment Simulation", Ocean Engineering, John Wiley and Sons, Inc. (Chapter 17), p. 633.

9. Beauchamp-Nobbs, E., "Naval Ship Research and Development Center's Ocean Pressure Laboratory", ASME Paper No. 68-WA/UNT-5.

10. Schuh, N. F., Jr., "Ocean Simulation Laboratory", ASME Paper No. 68-WA/UNT-2.

11. Schuh, N. F. and Jasper, N. H., "Ocean Simulation Laboratory", Southeast Section S.N.A.M.E., May 10, 1968.

12. Finlayson, L. A. and Cole, C. K., "Simulationg Deep Ocean Environments", Oceanology International, Sept./Oct. 1968, p. 40.

13. DeHart, R. C. and Briggs, E. M., "Southwest Research Institute Underwater Engineering Laboratory", ASME Paper No. 68-WA/UNT-6.

14. Allnutt, R. B., "Use and Design of Pressure Tanks for Deep Sea Simulation Facilities", ASME Paper No. 68-WA/UNT-4.

ABBREVIATIONS

For the reference listings which follow, the abbreviations used are given below. The Institution of Naval Architects became the Royal Institution of Naval Architects in 1960. The Journal of the American Society of Naval Engineers became the Naval Engineer's Journal in May 1962.

ABS	American Bureau of Shipping
AIAA	American Institute of Aeronautics and Astronautics
AIME	American Institute of Mining, Metallurgical and Petroleum Engineers
AISC	American Institute of Steel Construction
API	American Petroleum Institute
ASCE	American Society of Civil Engineers
ASME	American Society of Mechanical Engineers
ASNE	American Society of Naval Engineers
ASWEPS	Antisubmarine Weapons Environmental Prediction System (U.S. Navy)
INA	Institute of Naval Architects (Great Britain)
Izv. Geophys. Ser.	Izvestiia Geophysics Series (Izvestiia Seriia Geofizicheskaia) (Soviet Union)
MIT	Massachusetts Institute of Technology
MTS	Marine Technology Society
NCRE	Naval Construction and Research Establishment (Gt. Britain)
NECIES	North-East Institution of Engineers and Shipbuilders (Gt. Britain)

OECON	Offshore Exploration Conference
RILEM	Reunion Internationale des Laboratories d'Essais et de Recherches sur les Materiaux et les Constructions
RINA	Royal Institution of Naval Architects (Gt. Britain)
SAMPE	Society of Aerospace Material and Process Engineers
SESA	Society for Experimental Stress Analysis
SPE	Society of Petroleum Engineers
SNAME	Society of Naval Architects and Marine Engineers
WHOI	Woods Hole Oceanographic Institute